THE
RENDLESHAM FOREST
UFO MYSTERY

AND
PROJECT HONEY BADGER

First Published 2022

Copyright © George Wingfield 2022

Published by The Squeeze Press
An Imprint of Wooden Books Ltd
Red Brick Building, Glastonbury, BA6 9FT

A CIP catalogue record for this book is available
from the British Library

ISBN-13: 978-1-906069-23-0

Designed and typeset by Wooden Books Ltd, UK.
Printed and bound by Replika Press, India

www.woodenbooks.com

the
SQUEEZE
PRESS

THE
RENDLESHAM FOREST
UFO MYSTERY

AND PROJECT HONEY BADGER

GEORGE WINGFIELD

CONTENTS

PART I

PART II

A-10	Fairchild Republic A-10 Thunderbolt II aircraft (aka "Warthog")
ADS	Active Denial System
AOS	Adaptive Optics System
AFB	Air Force Base
AFOSI	Air Force Office of Special Investigations
ATP	Advanced Theoretical Physics Project
A1C	Airman 1st Class
APRO	Aerial Phenomena Research Organization
ARRS	67th Aerospace Rescue and Recovery Squadron (at RAF Woodbridge)
ASCII	American Standard Code for Information Interchange
BCWM	Bentwaters Cold War Museum
BP	Boilerplate (as for Apollo Command Module & other spacecraft)
BP CM	Boilerplate (Apollo) Command Module
C-141	Lockheed C-141 Starlifter military strategic transport aircraft
CAUS	Citizens Against UFO Secrecy
CCWC	(or CCW) the Convention on Certain Conventional Weapons (1980)
CH-47	Boeing Chinook helicopter
CIA	Central Intelligence Agency
CINCUSAFE	Commander-in-Chief US Air Force in Europe
CLI	Cash-Landrum Incident (my acronym)
CM	(Apollo) Command Module
CSETI	Center for the Study of ExtraTerrestrial Intelligence

CSC	Central Security Control at RAF Bentwaters/Woodbridge twin bases
CSM	(Apollo) Command & Service Module
DEW	Directed-energy Weapon
DIA	Defense Intelligence Agency (US Army)
EBCDIC	Extended Binary Coded Decimal Interchange Code
ERF	*Encounter in Rendlesham Forest* (2014 book)
ETA	Extraterrestrial Alien
ETH	Extraterrestrial Hypothesis
FMWC	False Multiple Witness Claim (my acronym)
FOI(A)	Freedom Of Information (Act)
HH-53	Sikorsky Super Jolly Green Giant & MH-53 Pave Low helicopters, also . . . Sikorsky RH-53D Sea Stallion helicopters
ICBM	Intercontinental Ballistic Missile
IRGC	Islamic Revolutionary Guard Corps
KSC	(John F) Kennedy Space Center, near Cape Canaveral, FL.
LANL	Los Alamos National Laboratories, in New Mexico
Laser	Light Amplification by Stimulated Emission of Radiation
LGS	Laser Guide Star
LE	Law Enforcement
LM	(Apollo) Lunar (Excursion) Module, (sometimes LEM)
LOT	Low Observable Technology
LPI	Laser Projected Illusion (my acronym)
MoD	(the UK's) Ministry of Defence

MSgt	Master Sergeant
MUFON	Mutual UFO Network
NASA	National Aeronautics and Space Administration
NATO	North Atlantic Treaty Organization
NOSS	Naval Ocean Surveillance System
NSA	National Security Agency
PHB	Project Honey Badger (1980)
PTSD	Post-traumatic Stress Disorder
RAF	Royal Air Force
RF	Radio Frequency
RFI	Rendlesham Forest Incident, Dec 26th 1980 (my acronym)
SAR	Search and Rescue
SAS	Special Air Service (UK)
SDI	Strategic Defense Initiative (aka "Star Wars")
SEALs	(United States Navy) Sea, Air, and Land Teams
SM	(Apollo) Service Module
SOAR(A)	160th Special Operations Aviation Regiment (Airborne)
SOF	Special Operations Forces
SOR	Starfire Optical Range (at Kirtland AFB, New Mexico)
SP	Security Police (As in 81TFW SP)
SSgt	Staff Sergeant
STOL	Short Take Off and Landing
TFW	Tactical Fighter Wing. (As in 81TFW)
THW	Trojan Horse Weapon (my acronym)
UAP	Unidentified Aerial Phenomenon (or, plural: Phenomena)
UAV	Unmanned Aerial Vehicle (drone)
UFO	Unidentified Flying Object

UFT / FT	(Unidentified) Flying Triangle (my acronym)
UH-60	Sikorsky Black Hawk helicopter (regular & LOT versions)
USAF	United States Air Force
USN	United States Navy
USSOCOM	US Special Operations Command
VA	(US Department of) Veterans Affairs
VLT	Very Large Telescope (Chile)
VR	Virtual Reality
VTOL	Vertical Take Off & Landing

Part I

Introduction

The title of this book, *The Rendlesham Forest UFO Mystery and Project Honey Badger*, may require a little explaining for those who are unfamiliar with the subject. There have been quite a few earlier books about the Rendlesham Forest UFO of December 1980 but none of these has linked what happened then with *Project Honey Badger* which was a secret operation by the US military to develop techniques for rescuing the American hostages held in Iran.

Since the extraordinary events in Rendlesham Forest in 1980 which involved a number of US Air Force (USAF) personnel based at the American controlled airfields RAF Bentwaters and RAF Woodbridge, the mystery of what actually happened there has been the subject of considerable controversy. The physical craft which undoubtedly descended into the forest was approached by two USAF SP airmen and has ever since been described as a "UFO"— Unidentified Flying Object. What the UFO really was is the main subject of this book.

What exactly does the description "UFO" imply in this context? One may well ask. Some might immediately infer that a UFO must be an alien spaceship from a different star system, even maybe from some remote part of our galaxy. Some might believe a mysterious UFO like this *must* be a craft from out of this world—perhaps even a spaceship piloted by humanoid aliens from beyond!

Although I still intend to use the term "UFO" to describe the craft which the US airmen encountered in the forest, I say there is now compelling evidence to show that the craft *was* of human origin and also just what its purpose was. It was *not* a "flying object" in the sense that it could land and take off under its own power. However, it was certainly intended that it should remain "unidentified" at the time it was placed in the forest and one reason for that was because nearly all its identifying markings had been deliberately erased.

The controversy over the Rendlesham Forest Incident (**RFI**) usually divides those who are interested in researching the enigma into three camps. First there are those USAF personnel who were involved with these events in December 1980, some of whom actually witnessed the UFO(s). Obviously we should listen to what they have to say and also try to evaluate whether they are being completely sincere as regards their recollection of these things. It's also relevant that some of these witnesses are still bound by security oaths which they took when they joined the US military over 40 years ago. If any of the men involved know, or think they know, the actual truth about these events, it may not correspond to the version of the story they see fit to tell us now.

The second camp among those who remain interested in the mystery comprises those who believe the UFO(s) seen there in December 1980 were indeed craft of extraterrestrial origin—whether or not any ET aliens were involved. I think it is fair to say that neither of the USAF men who closely approached the Rendlesham Forest UFO currently believes it was an alien spacecraft or, indeed, anything of non-human origin. Several other interested parties who did not serve in the US military have however written about or produced films and TV documentaries on the Rendlesham mystery. Some of them clearly do believe the UFO was of alien origin. They are mostly staunch UFO-believers and it's very likely they will dispute much of what I've got to say here. However I strongly believe this book is far closer to the truth than anything previously published about the Rendlesham mystery.

The third camp in this controversy is that band of UFO/ET skeptics who are simply not having any of it. When the Rendlesham Forest UFO mystery first became public knowledge certain prominent UFO skeptics proclaimed that this was all nonsense and those making such claims were either lying or were delusional. It was first suggested the US airmen had failed to recognize the rotating beam of Orfordness Lighthouse six miles away and it was that they had been chasing.

When that explanation proved inadequate it was suggested they had witnessed a meteor shower or the re-entry of a Russian satellite's booster rocket that supposedly descended into Earth's atmosphere at the very time in question. But what then was the actual physical craft which the airmen encountered in the forest? Most UFO skeptics insisted the men were lying and the story was contrived. Or else, these witnesses had been either drunk or drugged and had mistaken some perfectly normal object such as the lighthouse for this UFO. A whole list of theoretical explanations was served up by the skeptics during the next 40 years but few of them came anywhere near the truth.

This book does not seek to prove or disprove the reality of supposed alien UFOs. That is something I will leave to others. Although few scientists these days would deny there is a strong possibility of life, even advanced intelligent life, elsewhere in the universe, the real question when it comes to UFOs is whether any such extraterrestrial life is visiting, or ever visited, this planet. All such claims should be considered and judged on their merits but I'm unaware of any alleged ET UFO case which has stood up to scientific scrutiny. Obviously some supporters of ETH (the Extraterrestrial Hypothesis) of UFOs seized on the Rendlesham Forest Incident as proof of their case but I say there is absolutely no proof that it was anything extraterrestrial.

The truth is that the field of UFOlogy, both in the US and the UK, has over the last 50 years been thronged with fantasists, hoaxers, forgers, and quite a few UFO authors who declare that what is really their own fiction is actually fact. There are, of course, some serious researchers of the subject but I fear these seem to be outnumbered when it comes to UFO conferences, TV shows on UFOs, and videos on the internet. Unfortunately, many UFO documentaries and much of the UFO subject these days consists of "fake news". In the fake news category over the years there have been many false claims of alien contact and, indeed, claims of physical abduction of humans by ET aliens. None of the claims of alien contact has ever proved true—or ever advanced the slightest proof of the ET aliens' physical reality.

In the Rendlesham Forest UFO case of December 1980, something of an apparently extraordinary nature definitely did happen. There were several reliable witnesses, but, even so, it has taken years of research to uncover the truth of what must have actually occurred. This book presents an explanation of what the UFO in the forest over 40 years ago really was. And, although I believe unbiased readers will accept what I'm suggesting, I'm not quite so naïve as to think that some diehards with entrenched positions regarding the Rendlesham Forest UFO will concede that, or even alter their fixed opinions one bit.

A HYPOTHETICAL SCENARIO FORTY YEARS AGO

Imagine, if you can, the dilemma facing the leader of an armed gang of young revolutionaries and their political comrades in the Middle East in 1979/1980. The revolutionaries have captured a number of American hostages during a violent uprising against a US-backed regime which has now been overthrown. A large mansion on the outskirts of the city is being used as a prison for the US hostages and the gang's leader has complete authority over the guards who are holding them there. The hostages must be guarded on a 24/7 basis and his men do long shifts watching all of them by day and by night.

None of the hostages can be allowed to escape and no one apart from the revolutionaries can be allowed to enter the house. Some of them have suggested shooting the hated Americans but the leader has told them that is forbidden. The reason—which some seem to fail to understand—is that the hostages are a valuable bargaining chip that the revolutionary regime is determined to use for political purposes. The hostage situation will give the revolutionaries leverage to force the United States to repay billions of dollars that were supposedly sent abroad or plundered from the country. If any of the hostages are killed or allowed to escape their value to the new ruling regime would be lost. They must be kept in captivity and they must be kept alive.

As the leader sits alone, dozing on and off in the middle of the night, one of the guards comes to report that there is something suspicious outside and that he should come and take a look. From an upstairs window the guard indicates a flashing light that looks to be about a mile away in an open area of scrub well away from the road. Nothing else appears visible outside and there's no traffic on this road.

What is it? It flashes red and white intermittently and the light sometimes appears to turn off for several minutes. But it doesn't move. It's hard to say what shape the dark object behind the light really is though it seems about the size of a small truck. No sound is heard from out there and binoculars show nothing else near it or anywhere in that direction.

Deeply puzzled the leader keeps watching and looking at it through binoculars from time to time. Can it possibly be something connected with their imprisoning of the US hostages? If not, could it be some sort of criminal activity going on? Or has a small plane crashed out there in the desert? Could it be a genuine UFO like the one that was apparently observed and chased by a military jet near there in 1976?

If it's the beginning of an attempt by the Americans to rescue the hostages there's no sign of any commandos or paratroopers approaching. No sign of any aircraft in the night sky and nothing which appears to threaten the guards and their hostages.

After nearly an hour the flashing light out in the desert has not moved. Someone has to go out there to investigate. The leader knows that it should be him. He selects two of the guards to go with him and tells them both to bring their guns. If there's any sort of trouble, he says, they will riddle this thing with bullets!

The other guards must stay behind and guard the hostages. No hostages are to be shot unless there's a full scale assault by American commandos and they succeed in breaking into the house where the hostages are being held. Having told the other guards that, the leader grabs his AK-47, plus a 40-round magazine. He leads his men out towards the unidentified object that's sitting out there in the desert. Since he will remain in radio contact with them, what's to be done if he fails to return hasn't even been considered.

Unknown to the three men, they are just about to encounter a secret US Special Forces Trojan Horse Weapon (**THW**). Its deployment would have been prelude to a full scale second US attempt to rescue the Tehran hostages alive and with a minimum number of casualties.

THW is my unofficial acronym for Trojan Horse Weapon and something of the kind was going to be needed in 1980—after the failure of an earlier rescue mission—for the planned second hostage rescue attempt to stand any chance of success.Everyone is familiar with the story of the great wooden horse which the attacking Greek army left outside the gates of Troy during the Trojan War (c. 1200 BC). The Trojans were very puzzled as to what this thing was but, believing the Greek forces had sailed away from Troy, they took it inside the city anyway. Inside that wooden horse a number of Greek soldiers were concealed and, in the dead of night, they climbed out and opened the city gates to allow their returning comrades to rush in. The clever deception allowed the Greek army to invade and destroy the city of Troy and it was this ruse which brought the lengthy Trojan War to an end.

But, obviously, weapons of deception today would have to be far more devious and sophisticated than any wooden horse used by the Greeks to gain entry to the city of Troy. In the late 20th century a THW would hardly contain armed soldiers but it would more likely be a hi-tech drone, a Laser Projected Illusion (**LPI**), or even a holographic display which was intended to deceive and confuse an enemy. Also, it might perhaps contain a Directed-energy Weapon (DEW) that could be used to disable or neutralize enemy personnel who approached it.

THE RENDLESHAM FOREST UFO CONTROVERSY OF 1980

If one casts one's mind back to the year 1980 and the main events that were in the news at the time, one could ask whether one of these might be connected in any way with RFI—the Rendlesham Forest Incident. There was the volcanic eruption of Mount St Helens on May 18th and there was the shooting of John Lennon in New York City on December 8th. There was the beginning of the Iraq—Iran War in September when Saddam Hussein's forces invaded Iran and this war continued for eight years with an estimated over one million dead.

On a lighter note, 1980 saw the introduction of the Rubik Cube and the release of the movie *E.T.—the Extra-Terrestrial*. But this was also a time when Cold War tensions between the Soviet Union and the Western Allies rose to new heights. Lech Walesa's independent labor union, Solidarity, began the movement that forced the withdrawal of Soviet troops from Poland and this led eventually to the end of Soviet domination in Eastern Europe. Another Cold War development was the Soviet invasion of Afghanistan during the first few weeks of 1980. One consequence of the invasion was that US President Jimmy Carter announced, on March 21st, that the US would boycott the Olympic games scheduled to take place in Moscow that summer.

However, it wasn't any of the above events that was connected with the extraordinary UFO mystery that took place in Rendlesham Forest in December. In November 1979, a group of Iranian college students, who were supporters of the Iranian Revolution and its spiritual leader Ayatollah Khomeini, took over the US Embassy in Tehran and held 52 US diplomats and citizens hostage for 444 days until January 1981. President Carter called the hostage-taking an act of blackmail and the hostages "victims of terrorism and anarchy". Subsequent unsuccessful US attempts to free the hostages crippled Carter's presidency and further military plans to rescue them continued right up to his final days before leaving the White House. It now appears that RFI was the testing of an untried military technique which would have used a device like the "UFO" that was placed briefly in Rendlesham Forest on a dark winter's night in 1980.

The Rendlesham Forest UFO has for years been the subject of books, documentaries, TV programs, speculation informed and uninformed, and indeed the source of endless controversy. I believe that the real solution to this enigma is now evident and I hope that those who have studied those events will agree that what I suggest is the truth of the matter regarding what the "UFO" actually was.

By UFO I *solely* mean Unidentified Flying Object here and there can be no dispute that it's the appropriate term to use. Obviously some may *assume* that UFO means a spacecraft or aerial vehicle of extra-terrestrial origin but that's not what "UFO" is intended to mean here. There definitely was a physical object that appeared to be some kind of a craft—which landed or was placed in a forest clearing. It left behind quite definite traces of its presence on the ground. To the two USAF airmen who approached it closely late on a December night it was something unidentified—and it was certainly of unknown origin.

Reference is sometimes made to "military UFOs" which have been seen in a military context or near a military base like the Rendlesham Forest UFO. The description leaves open the question of whether such objects are actual physical craft or merely misperceptions. Some UFO researchers claim "military UFOs" are of extraterrestrial—i.e. alien—origin and must have been captured by the US military or else back-engineered from alien craft. This speculation seems highly unlikely as regards RFI and I hardly think one needs consider it here!

Certainly this book is not intended to debate the question whether some UFOs are really of extraterrestrial origin or not. Its purpose is solely to get at the truth and to propose a solution as to what the UFO which landed in Rendlesham Forest in December 1980 really was. It also examines the case of another very physical UFO seen up close in Texas during that same week and with striking parallels to RFI.

SOME FALSE EXPLANATIONS FOR WHAT THIS UFO WAS

Before we go any further it's worth saying that there are plenty of things that the UFO was quite definitely not. For a start it was *not* the Orfordness Lighthouse which certain pundits maintained was the real explanation for the strange lights which the USAF airmen had seen. In particular, astronomer and UFO skeptic Ian Ridpath and others insisted for years that the lighthouse's beam explained it all—though Ridpath later suggested it was a bright meteor which apparently blazed across the sky that night. Other UFO skeptics said that the fireball, if indeed one was seen in Suffolk, must have been space debris from Russian satellite Cosmos 749 re-entering the atmosphere. But that had happened at roughly 9:08 pm on December 25th—some hours before the Rendlesham UFO was seen in the forest.

Such false explanations probably assumed the USAF witnesses, two of whom closely approached—and one even touched—the landed UFO, were liars, morons, or else high on drugs perhaps hallucinating at the time. The primary witnesses did mention the lighthouse beacon in their reports of the incident but were adamant the UFO which landed in the forest was something completely different. It had colored lights on it, which they saw, and was definitely some kind of physical craft.

To which it must be added that anyone familiar with the area would have known soon enough what the beam of the 98 ft high Orfordness lighthouse was. I travelled to this part of Suffolk several times in the 1970s visiting Orford, Orford Castle, the town of Woodbridge and Sutton Hoo among other places. On a clear night the lighthouse's revolving beam produced a bright white flash every 5 seconds easily visible from anywhere within about ten miles— even if the lighthouse itself was not directly visible.

In 1980 the regularly flashing beacon was located about six miles due east of RAF Woodbridge's East Gate. It was from here the US airmen went out into the forest to investigate the landed UFO.

Orfordness Lighthouse today

Admittedly that was in the same direction as that where the UFO appeared to have landed but the lighthouse certainly did not explain what they were seeing. (Orfordness lighthouse was decommissioned in 2013 because of the encroaching sea. Unless demolished, it's likely to fall into the North Sea within the next ten years or so.)

Then, there was a BBC program about the Rendlesham Forest UFO in which another pundit told us that the flashing lights must have been those of a local police car with red or blue lights being driven along a forest track. Here was a further absurd suggestion that the USAF witnesses who were charged with the security of RAF Bentwaters and RAF Woodbridge, the twin bases where nuclear weapons were stored, must have been half-witted or else drunk and/ or drugged.

More silly explanations followed—like the one offered by an Ipswich man, Peter Turtill, aged 66, who claimed he had driven a stolen truck loaded with agricultural fertilizer, which had somehow caught fire, into the forest where he abandoned it. Then, there was another man who claimed when he was based at RAF Woodbridge he had taped colored plastic over his car headlights and had driven the car on the runway there to fool other USAF personnel that it was a flying saucer. That, he might well have done sometime but it hardly explained what the USAF witnesses had seen in Rendlesham Forest on the nights of December 25th/26th and 27th/28th, 1980.

In December 2018 the *Daily Telegraph,* once again, carried a story claiming to explain the Rendlesham Forest Incident. An unidentified person—calling himself "Frank"—who claimed to have once been in Britain's SAS told UFO researcher David Clarke the incident was a revenge prank on US airmen by SAS members who were "captured, beaten up, and brutally interrogated" after parachuting into the twin bases to test security. Black helium balloons and kites activated by remote control were supposedly used to simulate UFOs and fool the Americans who had come out into the forest to investigate.

It is completely unlikely that SAS men would ever have been

allowed to parachute unannounced into an active frontline USAF base in 1980 either by their own commanders or by the US top brass. The story was almost certainly fake news and had no evidential value whatever. Even UFO skeptic Clarke who consulted a former SAS member, Robin Horsfall, had to concede that the story was dubious in the extreme and that "Frank" did not sound as if he had ever been a SAS member.

Such dubious supposed explanations for the Rendlesham UFO were trotted out from time to time by serious newspapers like the *Daily Telegraph*, whose skeptical editorial staff seemed to be desperate to disparage anything UFO believers claimed regarding the RFI. Any suggestion the UFO was an alien spaceship was indeed almost certainly nonsense—but so too have been most of the alternative explanations that were put forward.

A CURIOUS EXPLANATION GIVES AN UNEXPECTED CLUE

There is one further explanation for the Rendlesham UFO incident that appeared in the *Daily Telegraph* on December 20th 2010 just a few days before the 30th anniversary of these events. When I saw the news item, which is reproduced below, once again I thought it was totally absurd like most of the others. However several years later, after more research, I suggest there's a clue here hiding in plain sight.

So, the Rendlesham UFO was now to be explained as a "moon pod" that had either been accidentally dropped in the forest by a USAF helicopter or perhaps put there as a prank to fool other US airmen! None of this sounded right at all. The *Daily Telegraph's* supposed "moon pod" must have meant an Apollo CM (Command Module) in which three-man Apollo crews had travelled to the moon in 1969 and the early 1970s. These CMs returned to Earth as the only recovered component of Apollo missions, carrying the

Britain's most famous UFO 'was moon pod'

By Daily Telegraph Reporter

BRITAIN'S most famous "alien" incident may have been caused by US airmen trying to cover up the fact they dropped an Apollo crew pod.

An incident that featured mysterious lights and the sighting of a strange object in the sky above a forest on Boxing Day 1980 has been termed "Britain's Roswell".

But as the 30th anniversary approaches, locals have suggested the real explanation lies with an American helicopter crew who bungled a transfer and tried to cover it up with the UFO claim.

They believe the "alien spacecraft" was actually the crew pod from an Apollo space rocket which was dropped in Rendlesham Forest, Woodbridge, Suffolk by a helicopter from the nearby US Air Force base.

Witnesses reported seeing bright fast moving lights in the sky leading to speculation that it was a UFO. Several badly-shaken American airmen gave detailed descriptions of the craft, and security teams guarding Nato nuclear weapons on the base raced to investigate, but the incident was allegedly hushed up.

Other explanations include the beam from a nearby lighthouse, a meteor shower or a Russian rocket. But Graham Haynes, a museum curator said: "One' couple said they saw a helicopter with a large cone-shaped object slung underneath it. It was flying low, probably hit the runway lights and dropped the capsule into the forest. They came back the next day to collect it. At the time the base was home to the team that were assigned to recover Apollo moon rocket capsules.'

A story in the Daily Telegraph of Dec 20th 2010 suggested that the Rendlesham UFO was an "Apollo moon pod" dropped in the forest.

crew, and splashed down in the Pacific Ocean. CMs to be recovered would separate from their LMs, Lunar (Excursion) Modules, and their SMs (Service Modules) before returning from the moon and re-entering Earth's atmosphere.

Most of those who have written about the Rendlesham UFO have taken the line that if it really was an Apollo CM that was dropped in the forest late on December 25th 1980 that must have been either a bona fide accident by the helicopter crew or else it was some kind of a silly jape by them to fool other US airmen at the twin bases.

Neither of those suggestions is true however and now there's every reason to believe it was placed in the forest by a helicopter of the 67th ARRS, then based at RAF Woodbridge, with a much more serious intention. I spoke recently with Graham Haynes, the Director of the Bentwaters Cold War Museum, who provided the *Daily Telegraph* with the story in the news item above. He knew about the couple who had seen a helicopter flying out of the Woodbridge base on Christmas night and who said a large cone-shaped object was slung below it.

One or both of them lived in Folly House which is situated very close to the east end of RAF Woodbridge's main runway and just outside the base perimeter fence. That was one of two semi-detached houses belonging to the Forestry Commission, who manage Rendlesham Forest. They let one of the houses to their employee, the forester. Anyone living in Folly House in 1980 would have been totally used to the deafening noise of jet aircraft and helicopters taking off or landing on that runway and, most times, might not have even stopped to look. But this was late on Christmas Day 1980 and there wasn't supposed to be any flying from the base that day. The couple who saw a helicopter with a large cone-shaped load slung below may even have recognized it as an Apollo CM since one of these had been flown into and out of the base at times in earlier years when the 67th ARRS practiced aerospace recovery operations in the 1960s and 1970s.

Only after reading about the story in the *Daily Telegraph* in 2010 did I discover that the boilerplate Apollo CM which had been kept by 67th ARRS at RAF Woodbridge for many years had actually been flown back to the USA in late 1977. It was the only Apollo CM that had ever been based in the UK and, if this was the CM seen slung below a helicopter in late December 1980, it was indeed a further mystery. In any case, it soon became apparent the very same boilerplate CM must have been returned to RAF Woodbridge—evidently under a cloak of secrecy and in mysterious circumstances—some time before the end of December 1980. That soon became plain since from after that time it was reported to be visible and located on the base. It could be even be seen from outside the base perimeter. The official line put out by 67th ARRS from early 1981 was that the boilerplate CM had never left!

THE UFO IS INVESTIGATED BY PENNISTON & BURROUGHS

Not long after midnight on the night of December 25th/26th Airman First Class (A1C) John Burroughs was one of the security police guards on duty who saw strange red and blue lights blinking in the forest. Burroughs had been patrolling at RAF Woodbridge and was near the East Gate of the airfield at the time. He and others who saw the lights assumed an aircraft might have crashed in the heavily wooded area to the east of the base known as Rendlesham Forest.

There was not meant to be any flight activity by USAF A-10 aircraft at the twin bases that night but there was a possibility maybe that some civilian airplane had come down in the forest. Burroughs drove out into the forest with SSgt Bud Steffens to investigate but after about 200 yards stopped when they again saw white, red and blue lights blinking ahead of them. At this point they decided to return to East Gate where they could report the situation by phone to airbase CSC.

It was now decided A1C Burroughs should investigate further but that he should be accompanied by the then on-duty Flight Chief at RAF Woodbridge, SSgt James (Jim) Penniston. The latter was requested to drive to the East Gate with his rider, A1C Edward Cabansag, and rendezvous with Burroughs and Steffens. It was then agreed that Penniston, Burroughs and Cabansag would proceed into the forest as before while Steffens remained behind at East Gate. He would take care of the security men's guns which they were supposed to leave behind whenever they went outside the base's perimeter.

As before, they drove as far as they could. They went eastward in the direction of the unexplained lights before stopping and proceeding on foot through the trees. They still were assuming that a small aircraft might have crashed in the forest despite the fact that Steffens had apparently said before they left him, "It didn't crash. It landed".

What happened next in the forest is perhaps best described in Jim Penniston's own words which he presented in a position statement to a press conference held at the National Press Club in Washington on November 12th 2007:

My name is James Penniston, United States Air Force Retired. In 1980 I was assigned to the largest Tactical Fighter Wing in the Air Force, RAF Woodbridge in England. I was senior security officer in charge of base security. At that time I held a top-secret US and NATO security clearance and was responsible for the protection of war-making resources for that base.

Shortly after midnight, on the twenty-sixth of December 1980, Staff Sergeant Steffens briefed me that some lights were seen in Rendlesham Forest, just outside the base. He informed me that whatever it was didn't crash . . . it landed. I discounted what he said and reported to the control center back at the base that we had a possible downed aircraft. I then ordered Airman Cabansag, A1C Burroughs to respond with me.

When we arrived near the suspected crash site it quickly became apparent that we were not dealing with a plane crash or anything else we'd ever responded to. There was a bright light emanating from an object on the forest floor. As we approached it on foot , a silhouetted triangular craft about nine feet long by six-point-five feet high came into view. The craft was fully intact sitting in a small clearing inside the woods.

As the three of us got closer to the craft we started experiencing problems with our radios. I then asked Cabansag to relay radio transmissions back to the control center. Burroughs and I proceeded towards the craft.

When we came up on the triangular shaped craft there were blue and yellow lights swirling around the exterior as though part of the surface and the air around us was electrically charged. We could feel it on our clothes, skin, and hair. Nothing in my training prepared me for what we were witnessing.

After ten minutes without any apparent aggression, I determined the craft was non-hostile to my team or to the base. Following security protocol, we completed a thorough on-site investigation, including a full physical examination of the craft. This included photographs, notebook entries, and radio relays through Airman Cabansag to the control center as required. On one side of the craft were symbols that measured about 3 inches high and two and a half feet across.

These symbols were pictorial in design; the largest symbol was a triangle which was centered in the middle of the others. These symbols were etched into the surface of the craft, which was warm to the touch and felt like metal.

The feeling I had during this encounter was no type of aircraft that I'd ever seen before.

After roughly forty-five minutes the light from the craft began to intensify. Burroughs and I then took a defensive position away from the craft as it lifted off the ground without any noise or air disturbance. It maneuvered through the trees and shot off at an unbelievable rate of speed. It was gone in the blink of an eye.

In my logbook (that I have right here) I wrote: "Speed—impossible". Over eighty Air Force personnel, all trained observers assigned to the Eighty-First Security police Squadron witnessed the takeoff.

The information requested during the investigation was reported through military channels. The team and witnesses were told to treat the investigation as "top secret" and no further discussion was allowed.

The photos we retrieved from the base lab (two rolls of thirty-five-millimeter) were apparently overexposed.

This account is obviously far more explicit than the first account attributed to Penniston after his interrogation by AFOSI agents not long after the actual event. In that unsigned and undated statement, with some of his sketches attached, he did recount approaching the colored lights which came from the object in the forest clearing. The object was *"definitely mechanical in nature"* but the closest to it he and Burroughs said they had approached was 50 metres (164 ft). When they proceeded further, Penniston said it "zigzagged" through the woods and was lost to sight.

Photo of James W. Penniston taken at National Press Club in Washington, D.C., on November 12th 2007

Clearly Penniston's earlier account had been sanitized by AFOSI and it might have been sanitized even more had they realized that at some stage it would be released to the public. Even so, AFOSI could hardly have been surprised when people, both on the base and off it, took the accounts of the object encountered in the forest to have been the popular conception of a UFO—one of the ET variety! Not that USAF officials discouraged such an explanation at the time, since it was much better for the general public to think "ET UFO" than uncover the real explanation which was a closely guarded secret.

At this stage one might well ask whether John Burroughs confirms Penniston's account of what actually happened after they had both approached to within 50 meters of the mysterious craft and then moved closer. The answer is that Burroughs has no recollection after that and there is no suggestion that he actually reached the landed UFO, touched it, and saw symbols on its surface as Penniston claims he did. Nor does Burroughs have any recall of the UFO silently taking off, rising slowly above the trees, and then vanishing into the sky at *"speed—impossible."*

The apparent reason Burroughs has no memory of the UFO itself is most likely because he was struck by what seemed to be a *"silent explosion of light"* from the UFO when the two men moved forward. Penniston says he looked to his right and saw Burroughs engulfed in a huge beam of light. He says both of them hit the ground instinctively and when he stood up again that was when he first saw a small metallic craft where a sphere of bright light had previously been.

If Burroughs was hit by an intense beam of electromagnetic energy it does seem to have concussed him and/or temporarily reduced him to a zombie-like state of which he has no recollection. We can certainly speculate what sort of a beam this could have been later. Penniston says that he next saw Burroughs standing in the clearing motionless. He yelled at him but got no reaction, and Burroughs did not move. Apparently he was unaware of Penniston yelling at him.

Since Burroughs has no recall of any of what followed, it is hardly surprising that the two men's accounts differ from this point on. That does not mean either account is necessarily untrue but it removes any independent confirmation of Penniston's claim of closely examining the UFO and its subsequent take-off. He says he then focused on the craft itself and the security implications for the twin USAF bases.

In the version of his story given to the National Press Club in Washington in November 2007 Penniston made no mention of Burroughs being incapacitated by a bright flash from the craft and it was certainly implied that both men went right up to the triangular UFO. *"After ten minutes without any apparent aggression"*, Penniston concluded *"the craft was non-hostile"* and *"we completed a thorough on-site investigation"*. The investigation, which included touching the craft and the symbols on it, he said had taken 45 minutes.

Well, now it definitely seems that, despite the "we", Burroughs took no part in the on-site investigation of the UFO so one has to wonder what became of him during that time. Was he standing motionless like a statue or had he run back behind the trees? Even if we accept Penniston's suggestion that "missing time" was involved in his alleged 45 minute inspection of the craft, there's an oddly uncomfortable disconnect between the two men's accounts of what happened.

WHAT HAPPENED AFTER ENCOUNTER WITH LANDED UFO?

After the UFO had supposedly vanished into the sky at unbelievable speed, according to Penniston, it seems Burroughs then reappeared and pointed the way out of the forest towards the east. He said they should carry on walking in that direction as moving lights were visible in the sky. Both men, and Cabansag who had rejoined them, walked on eastward across the fields for a further two miles according to Burroughs. Penniston says that they had further sightings of strange lights on the horizon and that at one stage *the object was so close that we thought it would land again*.

What they saw contrasts somewhat with the initial statement made by A1C Edward Cabansag who had retreated from the edge of that forest clearing while Penniston and Burroughs had gone ahead towards the UFO. He said that when the blue and red lights from the UFO in the forest were no longer visible, one could still see the yellow beacon light out beyond the forest. By "beacon light" Cabansag evidently meant the rotating beam of Orfordness Lighthouse which would have swept across the sky every five seconds even though the lighthouse itself was not directly visible. From the clearing where the UFO had been the lighthouse would have been a further five miles to the east.

Cabansag said the three of them had walked a good two miles out from where they had left their vehicle, going directly east beyond the edge of the forest. They saw a glow in the direction of the beacon light but as they got close to it they realized it was a lit up farmhouse (Green Farm at Capel Green). The farmhouse is situated across a field just 400 yards from where the UFO in the clearing had been. Burroughs mentioned in his initial report that when crossing this field before reaching the UFO they heard strange noises, like a woman screaming. They also heard farm animals—presumably at Green Farm—making a lot of noise. There was a lot of movement in the woods, which were seemingly lit up by light from the landed UFO. MSgt J.D.Chandler also mentions the noise of farm animals running around in his report of the incident. It is certainly possible the UFO in the forest had disturbed animals at the farm but these reported noises are odd since there weren't meant to be any pigs or cattle at Green Farm then. Nor does it appear any researcher of the UFO incident asked the farmer what he heard or saw at that time.

If the men did in fact walk a further two miles east they would have reached the Butley river (which flows north/south) and would have had difficulty crossing it. When they realized that the "beacon light" was further off than they had thought, they informed CSC by radio and Lieutenant Fred Buran told them to terminate the investigation. A1C Cabansag himself made no claim of seeing a triangular craft in the forest clearing or of seeing such a craft take off and shoot away.

While the initial reports submitted by Buran, Penniston, Burroughs, Cabansag and Chandler were almost certainly "sanitized" by AFOSI interrogators to remove any description of what the brightly lit mechanical object in the forest really was, it doesn't seem there was an attempt to falsify the timescale of the encounter and its aftermath.

The approximate timescale for the Rendlesham Forest Incident which happened during the early hours of 26th December 1980 appears to be like this:-

Above: A recent photo (looking W) of East Gate at RAF Woodbridge, the former NATO/USAF base outside which the RFI events occurred. A mile E of here, in the forest, a UFO landed on the night of Dec 25/26 1980

03:00 Penniston and Burroughs set off from RAF Woodbridge in a Security Police vehicle to investigate the strange lights. Also Cabansag & Chandler. They drive on road out of East Gate.

03:05 Vehicle stopped after ~500 yards and they proceed into the forest on foot walking through the trees for about ¾ mile. Chandler stays with vehicle to relay radio contact to CSC.

03:20 Arrive at forest clearing where UFO sits with its bright lights changing in no set pattern. When Penniston & Burroughs go towards the UFO it emits a huge flash of light towards them.Penniston reaches the object, touches it, walks round it and carries out careful examination supposedly lasting 45 min.

03:?? Penniston says UFO silently lifts off, weaves through trees & flies away at "speed—impossible." Burroughs reappears and leads the way out of forest toward east. With Cabansag, they walk further one/two miles that would have taken ~30 mins.

03:43 Is the time that Lieut. Fred Buran at CSC gives them orders by radio to terminate their investigation and return to base.

04:35 Approx time of return of men to Woodbridge & Bentwaters.

There is no reason to think that this timescale was incorrect and, if right, it's clear there was absolutely insufficient time for Penniston's alleged 45 minute close inspection of the landed UFO. What possible explanation can there be then for his conflicting account?

In the book *Encounter in Rendlesham Forest* Penniston offers what he says is the most likely explanation:

> "I suppose that anything is possible with this time discrepancy. I believe it is more than likely that within the affected area around the craft there was a [time] distortion of some kind, which based on the missing time from our watches indicates this, by them running forty-five minutes slow. We were definitely affected by this phenomenon in a physical way, including the machinery we wore (watches)".

This too, apparently, is claimed as John Burroughs's explanation for the missing time and he also is said to claim that shift commander (Fred Buran) at CSC had confirmed the men were missing for 45 minutes.

One concept of "missing time" is that sometimes found in UFO literature where supposed UFO abductees—or those who say they remember having a close encounter with a UFO—later discover that they've been away, as regards real time, for much longer than they originally thought. Either that, or like here, they claim to remember some otherworldly experience that seems to have taken much longer than the time they were actually absent or missing.

Needless to say, such claims are controversial and most usually not accepted by conventional science. There would usually not be any conclusive evidence of missing time from a slow or stopped timepiece. That proves very little. In the Rendlesham Forest case Fred Buran's confirmation that the men were missing for 45 minutes is a quite different from supposed "missing time". Buran was in fact stating Penniston and Burroughs had been out of radio contact for about 45 minutes. That presumably was because their radio contact with CSC (via Cabansag and/or Chandler) had failed or hadn't been attempted. It could have been because both of them were distracted or busy.

An alternative explanation for this missing time is that Penniston's story of the encounter is not 100% factual or else was embellished—either consciously or subconsciously—to fit the expectations of the UFO community. The idea that the UFO was some kind of alien craft that landed in the forest and later flew off, utilizing whatever secret motive power it possessed, is clearly one that would appeal to them.

In suggesting this, I don't say the whole story of the encounter in the forest is untrue or that Penniston didn't get close up and personal with the UFO—even to the extent of saying he received a download of binary digits from it. There is much to suggest he did these things but it may well have taken less than the 45 minutes which he claims. If John Burroughs was zapped and incapacitated when approaching the UFO, surely most people, quite unlike a superhero, would have turned and fled? Penniston knew that his duty was to photograph and check what this thing really was—despite the terror which that might engender—but most men would have done so as quickly as possible to minimize the risk.

I doubt if either man had the slightest idea what this thing was when they first approached it. Both were responsible, patriotic

American airmen with the 81ˢᵗ Security Police Squadron and they knew their duty was to guard the security of the twin bases against intrusion or attack. They had initially thought it might be a light aircraft which had crashed in the forest but, when they reached it, clearly it was not. SSgt Penniston soon realized the UFO was some kind of craft but to him its origin was completely unknown. Quite soon he probably considered two possibilities: (1) that it was either a secret military aerial vehicle or, perhaps, (2) a UFO of unknown origin. As we shall see, both men were relentlessly interrogated by AFOSI agents (or, more likely, men from the CIA or NSA) just 3 days later and they found themselves being treated like enemy combatants.

Undoubtedly each of their lives was severely changed by the UFO encounter experience, both psychologically and even physically. At some stage, perhaps back then or perhaps years later, each of them may have begun to realize they had been used as guinea pigs in a secret experiment by the US military. If they really thought that, neither could ever admit it or speak out against the perpetrators— probably not even today. Still bound by their military oaths and considerations of secrecy it is unlikely that will ever happen. What each of them now really believes the UFO in the forest was is probably a secret they will take to their graves.

PHYSICAL TRACES LEFT BEHIND BY THE UFO IN THE FOREST

When Penniston, Burroughs and Cabansag were ordered back to base by Lieutenant Fred Buran at CSC at 0343 hours on 26ᵗʰ December 1980, they walked back to the forest more or less the way they had come. Burroughs and Cabansag then walked together on a forest track whereas Penniston says he went straight through the trees to the small clearing where the UFO had been. He says that was when he first saw the three indentations in the ground which had

Above: Satellite view of Woodbridge Airfield (formerly USAF base RAF Woodbridge) showing the main runway. Positions of East Gate, Folly House and Green Farm are also marked. The small clearing in the forest where the UFO was located is almost exactly one mile from East Gate which gives scale to this map

Scale: 1 mile
(satellite ground image by Apple Maps)

been left there by the UFO. Some accounts say that Burroughs also saw these at that time but he only remembers going to see the marks later that morning well after daybreak.

When the men finally got back to CSC at Bentwaters they reported to Lieutenant Fred Buran on what had gone on in the forest. Incident reports were filled out in accordance with normal law enforcement (LE) practice at the twin bases. Soon after Burroughs and Penniston and the others went off duty Colonel Charles I. Halt, the Deputy Base Commander, arrived at CSC and asked what was going on. He was told by the duty desk sergeant, "Crash" McCabe that Penniston and Burroughs had been out during the night "chasing UFOs."

Colonel Halt, who was shortly to become a central figure in the Rendlesham UFO saga, expressed surprise but told McCabe to put everything down in the LE blotters—meaning the logs in which everything of significance that had happened during the shift was reported. That had been done, but later that day Halt realized that when he had suggested putting the more acceptable expression "unexplained lights" rather than "UFOs" in the blotters it hadn't been implemented. Soon afterwards he found that the LE blotters and the incident report had all been pulled and classified as SECRET by Base Commander Ted Conrad, Halt's immediate superior.

One obvious interpretation of Conrad's move was that he knew about a secret operation that had been carried out in Rendlesham Forest during the night and had taken the initiative to contain anything about it getting out. Certainly the motive for removing the blotters was containment but that failed to happen and word soon got around on the base leading to wild speculation that a UFO had landed.

Major Edward Drury was Deputy Squadron Commander to the more senior Major Malcolm Zickler. The latter was in command of the 81st Security Police and the Law Enforcement squadron. Drury had been woken up and briefed by Lieutenant Fred Buran who had also notified Zickler and Conrad. The on-duty shift commander on that morning was Captain Mike Verrano who had been among the first to question Penniston and Burroughs. Drury and Verrano decided they should visit the supposed UFO landing site themselves and arranged for Burroughs and Penniston to rendezvous with them. Master Sergeant Ray Gulyas was also told to attend and take photographs.

Burroughs and Penniston arrived at the forest landing site first and immediately saw three circular indentations exactly where they had seen the UFO during the night before. These were quite distinctive and each was 1½" deep and 7" diameter. The three markings formed an equilateral triangle with 8 ft sides. At a much later stage some of the UFO skeptics and others wanting to disparage the story would insist that the markings were anything but distinct and were most likely rabbit scratchings in the forest's soil.

The falsity of that suggestion was demonstrated by the fact that some of those who went out to look at the indentations that day took plaster casts of the marks which showed they were round and symmetrical. Drury, Verrano and Gulyas arrived soon after Burroughs and Penniston and the latter showed them the three ground markings. He described once more what he had seen in the forest clearing before leaving to return to his lodgings in Ipswich and get some rest.

But Penniston was unable to rest and decided that he should get further proof—if just for his own peace of mind—that this very real physical UFO had actually been standing there in the forest. He went back into the forest taking with him in a knapsack some plaster of Paris, a jug of water and a small bucket. When he reached the landing site he mixed the plaster and water placing some in each of the three indentations to make casts. He waited for about an hour while the plaster set before removing them, carefully wrapping them in plastic and putting them in his knapsack.

As he was leaving the forest he ran into Drury, Verrano and Gulyas who were on their way back to the landing site with a British police officer to show him the indentations. Drury asked Penniston why he'd returned to the site and said he should go home and get some rest. Penniston said nothing about the plaster casts he'd made. It would be his secret confirmation of the landing for many years after.

With the others Gulyas examined the site again, taking measurements and shooting a complete roll of film. He later handed this to Verrano who took it to the base photo lab to be developed. Subsequently he was told by Verrano all of his photos had been fogged and had been discarded. In fact, every one of the photos sent to the base lab during the UFO saga was pronounced fogged and worthless and it seems very likely all were confiscated or destroyed to keep the affair secret.

At a later stage Gulyas, still very puzzled, returned to the landing site alone and took his own black and white photos. Some of these survive though they are of poor quality. He too decided to make plaster casts of the indentations so that he'd have definite proof that the heavy UFO—whatever it was—had actually stood there on the hard ground. Rabbit scratchings? I hardly think so!

There were additional physical traces left behind by the Rendlesham Forest UFO which were investigated a day later and which left those involved in little doubt that a heavy craft of some description had descended into the small forest clearing. Some of the pine trees there had been impacted and abraded well above ground level and on the sides facing the center of the clearing where the ground indentations were. Samples of sap, which exuded on the trees where the abrasions were, were taken though it is not known if these showed anything abnormal. It seems the investigators assumed these abrasions must have been caused by a solid heavy craft either taking off or landing under its own power.

In retrospect, one now sees a more likely explanation was that the "UFO"—this heavy craft—dangled below a helicopter and caused the abrasions when it struck some of the trees at the edges of the clearing. That could have happened when it was lowered vertically into the clearing but is more likely to have occurred when it was later removed by the same helicopter. According to a forester named as James Brownlea in Jenny Randles' book *UFO Crash Landing* there was even evidence of a heavy object "having been dragged along the ground to remove it from the area". Unless this "UFO" was lifted up vertically, it may well have made some drag marks before leaving the ground.

Quite apart from the tree abrasions, the investigators took a Geiger counter with them and detected higher than normal radiation levels in the very position where the UFO had rested. Although there was no question that some craft of a very unusual nature had descended into the forest, and had later disappeared, years later UFO skeptics and, indeed, some of the US military figures involved tried to dismiss any such suggestion. In particular the notorious Halt Memo of 13th January 1981—which we shall come to—leaves one in no doubt about this "UFO" visitation and its physical reality. For many years now the naysayers have denied it and imply that Halt and the other USAF men were liars, fraudsters or fantasists. They felt quite sure that ET UFOs did not, and could not, exist. Or—as regards some of the USAF officers—they knew what this extraordinary episode really had been and realized it was a military secret, so the truth could never be told.

A KEY RENDLESHAM WITNESS WHO SEEMS TO BE SINCERE

Deputy Base Commander Charles Halt has consistently told his story of going out with some of the other airmen into the forest on the night of December 27th/28th and seeing a bright red or orange oval object with a black center—that reminded him of an eye. This was seen to maneuver horizontally through the trees, receding from the men and its bright light changed into five white lights which soon disappeared. Later several objects with multiple lights were seen in the northern sky and others were seen to the south. This UFO display continued for a period of 2-3 hours. One object apparently sent a concentrated beam of light down not far in front of Halt's feet and it was also said to have flown close to the Weapons Storage Area at RAF Bentwaters.

The encounter by two USAF security guards in Rendlesham Forest with multiple strange lights that were brightly colored was described in Halt's original memo of 13th January 1981 and similar events are also described on a tape recording which he made at the

Colonel Charles I Halt

DEPARTMENT OF THE AIR FORCE
HEADQUARTERS 81ST COMBAT SUPPORT GROUP (USAFE)
APO NEW YORK 09755

REPLY TO
ATTN OF: CD

13 Jan 81

SUBJECT: Unexplained Lights

TO: RAF/CC

1. Early in the morning of 27 Dec 80 (approximately 0300L), two USAF security police patrolmen saw unusual lights outside the back gate at RAF Woodbridge. Thinking an aircraft might have crashed or been forced down, they called for permission to go outside the gate to investigate. The on-duty flight chief responded and allowed three patrolmen to proceed on foot. The individuals reported seeing a strange glowing object in the forest. The object was described as being metallic in appearance and triangular in shape, approximately two to three meters across the base and approximately two meters high. It illuminated the entire forest with a white light. The object itself had a pulsing red light on top and a bank(s) of blue lights underneath. The object was hovering or on legs. As the patrolmen approached the object, it maneuvered through the trees and disappeared. At this time the animals on a nearby farm went into a frenzy. The object was briefly sighted approximately an hour later near the back gate.

2. The next day, three depressions 1 1/2" deep and 7" in diameter were found where the object had been sighted on the ground. The following night (29 Dec 80) the area was checked for radiation. Beta/gamma readings of 0.1 milliroentgens were recorded with peak readings in the three depressions and near the center of the triangle formed by the depressions. A nearby tree had moderate (.05-.07) readings on the side of the tree toward the depressions.

3. Later in the night a red sun-like light was seen through the trees. It moved about and pulsed. At one point it appeared to throw off glowing particles and then broke into five separate white objects and then disappeared. Immediately thereafter, three star-like objects were noticed in the sky, two objects to the north and one to the south, all of which were about 10° off the horizon. The objects moved rapidly in sharp angular movements and displayed red, green and blue lights. The objects to the north appeared to be elliptical through an 8-12 power lens. They then turned to full circles. The objects to the north remained in the sky for an hour or more. The object to the south was visible for two or three hours and beamed down a stream of light from time to time. Numerous individuals, including the undersigned, witnessed the activities in paragraphs 2 and 3.

CHARLES I. HALT, Lt Col, USAF
Deputy Base Commander

Above: Lieutenant Colonel Charles I. Halt's official report on the landed UFO in Rendlesham Forest at 0300 on December 26th 1980 + the subsequent events and sighting of the UFOs. (Note date errors in report)

time. He stated that he had never before seen anything like what he witnessed that night. "*I have no idea what we saw but do know that whatever we saw was under intelligent control*".

In a notarized affidavit made in June 2000 he concluded with his assessment that the objects he saw at close quarters were extra-terrestrial in origin and that the security services of both the US and the UK have attempted—both then and now—to subvert the significance of what happened at Rendlesham Forest and RAF Bentwaters by the use of well-practiced methods of disinformation.

I believe that Charles Halt is sincere in what he has told us and that indeed the USAF authorities covered up what really occurred by use of disinformation. However I would suggest there is an alternative explanation to this night-time UFO display being of extraterrestrial origin. It seems much more likely to have been a spectacular light display using lasers that was put on by US Special Ops men who had concealed themselves in the forest.

In any case there was no proof that any physical craft had descended into the forest on the night of December 27th/28th as one definitely had done about 48 hours earlier. If one accepts the first was a secret military test, one might assume this second apparent UFO visitation was a further exercise to develop techniques intended to distract and confuse an enemy. Whether or not such techniques were intended to be of use in a possible hostage rescue operation is debatable, though fake UFOs could well be considered an alternative variety of THW.

COLONEL HALT AND HIS TEAM SENT OUT TO INVESTIGATE

Word about Penniston and Burroughs's UFO encounter in the forest soon got around on the twin bases. Some thought it must have been a joke of sorts but others speculated that it could have been a genuine UFO of unknown, perhaps alien, origin. Wild stories of UFOs would soon circulate, as they did later among some of the civilian population with an interest in such matters who lived in the nearby area.

It was claimed that there were further UFO sightings on the night of December 26th by different members of the Security Police who were then on duty. A fiery red/orange object surrounded by a blue and white corona was said to have descended into the forest. Red, green and white lights were seen to appear and disappear among the forest trees. Beams of light were reported to have struck one of the security guard vehicles and a blue light said to have gone right through it.

It is difficult to verify some of these UFO stories all these years later and one could suspect they had been exaggerated and embellished if indeed they had much substance in the first place. Lights in the forest and lights in the sky maybe—but there was nothing to prove there had been any further visitation by an unknown craft or a heavy device that left physical traces on the ground like those on the first night.

We then get to events on the night of December 27th/28th when Col. Halt and three of his men went out into the forest once more to find what was going on. Halt was attending an awards dinner for members of the Combat Support Group in Woody's Bar on the base at RAF Woodbridge. At a late stage during the social function Lieutenant Bruce Englund, who was the shift commander then on duty, arrived looking white as a sheet and announced "The UFO is back". We are not told who had seen what, but must assume lights had been seen in the forest again in much the same direction as before.

Senior officer at the awards dinner was Base Commander Colonel Ted Conrad who was Halt's immediate superior. The two men conferred on what to do. It seems with hindsight that Conrad may have been one of the very few people at the twin bases who had any inkling of what was really going on. It had been on his orders that the blotters and all reports of Penniston and Burroughs's encounter in the forest had been pulled and he was treating the whole UFO matter as highly classified. Conrad's decision was that he would remain at the social function and it should be Halt who went out off the base and into the forest to investigate whatever this was.

One might have expected Halt and his companions to have driven immediately to a position on the Woodbridge base from which they could see any lights reported in the forest and assess the situation. However, there was no mention of this being done and it seems their first objective was to collect light-alls which might allow them to see anything that was happening on the ground on this cold winter night. A light-all is the name given to a wheeled platform with large lamps mounted on it. These are powered by a generator driven by a gasoline engine and the light-alls were available for general use on the base for such purposes as illuminating the night-time refueling of aircraft.

Besides Bruce Englund, Halt took with him Sergeant Monroe Nevels who was an experienced photographer. Besides his camera Nevels also brought a Geiger counter to measure any radiation at the UFO landing site. This site was their prime target at which the Geiger counter was certainly put to use. However, it's not recorded whether Nevels tried to photograph any of the lights seen in the forest or even those later seen in the sky. Halt himself took a mini-cassette tape recorder with several spare batteries and cassettes. They also took flashlights, radios and a night-vision device. The fourth member of Halt's team was Master Sgt. Bobby Ball, the on-duty flight chief.

Halt wanted several light-alls to be taken into the forest but at first it was found that most were not working. So, many needed refilling with gas and the high demand led to scenes of confusion at Woodbridge's military gas pumps. It appears there were several other vehicles there in addition to those with the light-alls ordered by Halt. Word had evidently got around at the twin bases that there was to be some sort of UFO hunt in the forest and there must have been several men from the two bases who wanted to join in out of curiosity, whether invited or not. In addition to Halt's small team of four, it's suggested that maybe as many as thirty or more men went out there on their own initiative during the early hours of December 28ᵗʰ. These certainly included John Burroughs and Sergeant Adrian Bustinza.

Halt and his men went first to where the UFO had supposedly landed some 48 hours earlier. He says he went out there determined to prove it was all nonsense but his initial skepticism evidently soon faded. They found the equilateral triangle formed by three distinctive indentations in the hard frozen ground. These the men had been told about, but they had never seen previously. It was plain that a large object had rested here on three legs, probably legs with landing pads, and must have weighed several tons. They next proceeded to check the site with their APN-27 Geiger counter.

There were high radiation readings in the indentations and also at the center of the triangle formed by them. Not abnormally high, but quite well above the general background level of radiation outside the actual landing position. A number of soil samples were collected for later analysis. A Geiger counter measures ionizing radiation such as alpha particles, beta particles, gamma rays and X-rays but that didn't mean there was necessarily any radioactive material left behind at the site. Non-ionizing radiation, say, from a powerful DEW, could have raised radiation levels temporarily at a site where it was discharged.

Years later UFO skeptics and some of the USAF officers from the base would claim the Geiger counter readings were insignificant or false. Their implication was certainly that the men were incompetent or else had never learned to use a Geiger counter properly. Their readings, it was suggested, were simply readings of normal background radiation and had nothing to do with any alleged UFO. These skeptics were most likely the same ones who maintained the symmetrical ground markings were no more than rabbit scratchings!

Charles Halt used his mini-cassette recorder to document what his team found at the UFO landing site and identification of radiation hot-spots there. Next they turned their attention to the pine trees surrounding the clearing. A few of these were slightly damaged, roughly at the 20 ft level, and they manage to take a sap sample from one tree for later analysis. The tape recorder was stopped and restarted as required in order to record these activities.

For quite some time then there was no word of any lights being seen in the forest or in the sky. The team were aware however of other people out in the forest and these included John Burroughs who made a request to join Halt's group via the CSC. The light-alls were eventually brought to the UFO landing site where Halt's team needed them but there was no mention on his tape recording of these lights revealing anything that was not visible otherwise.

WHAT HALT & HIS MEN CLAIMED THEY SAW IN THE FOREST

At about 3 am Halt's team saw, directly to the east, an unusual red sun-like light *"like an eye—oval-shaped, glowing and with a black center"*. It was 10 to 15 ft off the ground and was moving through the trees. Beyond the original UFO landing site was a barbed wire fence, the farmer's field, farmhouse and a barn. (This must have been Green Farm in the hamlet of Capel St Andrew.) Halt said the farm animals were making a lot of noise—presumably upset by this *"strange small red light"* that was perhaps half a mile away. It was due east of where Halt was, in the same direction as Orfordness lighthouse, but the red light could not have been confused with the rotating lighthouse beam.

Halt went on to say they ran towards this red light stopping only at the barbed wire fence.

"As it moved over the field it appeared to be dripping what looked like molten steel out of a crucible, as if gravity were somehow pulling it down. Suddenly it exploded—not a loud bang, just booompf,

and broke into five white objects that scattered in the sky. Everything except our radios seemed to return to normal".

(If any "molten steel" did fall in the field, certainly no metallic residue was found, as was untruly claimed by one UFO fan many years later!)

Halt's narrative continued as follows:

"We went to the end of the farmer's property to get a different perspective. In the north, maybe 20 degrees off the horizon, we saw three white objects—elliptical like a quarter moon but larger—with blue, green and red lights on them, making sharp angular movements. The objects eventually turned from elliptical to round".

He then says: *"I called the command post, asked them to call Eastern Radar, responsible for air defense of that sector. Twice they reported they didn't see anything* [on the radar]". (Like many UFO reports, nothing was seen on radar implying the UFOs were not physical craft but more likely moving lights in the sky perhaps produced by lasers.)

"Suddenly from the south a different glowing object moved towards us at a high rate of speed, came within several hundred feet and then stopped. A pencil-like beam, six to eight inches in diameter shot from this thing right down to our feet. Seconds later the object rose and disappeared."

There was no doubt, as he has repeated many times, that the objects he and the others saw were being intelligently controlled. "I have no idea what we saw but do know whatever we saw was under intelligent control."

Years later in 2010 Halt stated that this pencil-like beam illuminated the ground about ten feet from him and *"we just stood there in awe."* The object then moved *"<u>back</u> towards Bentwaters"*

(which was to his north, not his south!) *"and it continued to send down beams of light, at one point near the Weapons Storage Area."* (Since RAF Bentwaters base and its Weapons Storage Area were out of sight behind the forest trees and 2½ miles distant, there may well be some confusion here.)

> *"The objects in the north were still dancing in the sky. After an hour or so I finally made the call to go in. We left those things out there. The film turned out to be fogged; nothing came out. But a staff sergeant later made castings of the indentations and I had the soil sample."*

Whether the castings to which he refers were those taken on December 26th by MSgt Ray Gulyas or whether further castings were made isn't clear. Also, it's not clear whether *all* of the film shot by Nevels at the UFO landing site and, possibly, that of luminous objects they saw later in the sky was fogged. If the film was sent to the base photo lab for development, it would most probably have been confiscated and deemed fogged—whether that was the case or not.

DID SIGHTINGS SHOW UFO RETURNED OR WERE THESE LPI?

The various objects which Halt says they saw in the sky could well have been laser beam projections—or LPIs as I refer to them in a later chapter. Most false UFOs produced in this way are silent and ghost-like, moving in the sky like some projected luminous image. Harder to explain is what Halt described as a *"pencil-like beam shooting right down at our feet for perhaps a second or two"*. However, that is certainly a possible effect which could be produced by a laser beam.

In describing LPIs I have suggested that thin pencil-like beams which appear to come down from above could also be other laser beams going upwards from the ground. If a further laser situated next to the source of a LPI "UFO" was turned on, it would be impossible for an observer to distinguish between a beam going up to a point in the sky and one appearing to come down from that point. If the observer is already seeing a moving LPI "UFO", which he believes to be a physical object in the sky, then the illusion is compounded and he'd assume the thin laser beam was coming down from that object. Laser beams do not diverge, like old fashioned searchlights, making it difficult to know whether a laser source is above or below.

Various light illusions can be created using lasers and these probably would have been totally unfamiliar to most people in 1980. Further accounts of beams of light shooting down to the ground from UFOs in the sky are found in the mid-1980s, notably during the UFO wave in Belgium at that time. On the night of December 27th/28th 1980—and possibly on a subsequent night—there were accounts of pencil-like beams of light shooting down from UFOs into the Weapons Storage Area at RAF Bentwaters. These accounts should be treated with great caution. Since these highly secure underground shelters held nuclear weapons at the time, the claims are certainly sensational but they are also impossible to verify. In any case, it would have been impossible, as I've said, to see RAF Bentwaters or its Weapons Storage Area from where Halt was in Rendlesham Forest about 2½ miles distant. Even out in the field near Green Farm it was simply not possible to see RAF Bentwaters as it would have been screened from view by forest trees.

Colonel Halt recorded his impressions of the physical traces he saw in the forest and later the objects he saw in the sky on his mini-cassette tape recorder as the events unfolded. The voices of others in his team are also heard, sometimes sounding panicked, but there is never any external sound, not that the UFOs they supposedly saw made any. At one point Halt says "We're hearing very strange sounds out of the farmer's barnyard animals . . . . They're very, very active, making an awful lot of noise." Curiously none of that noise is on the tape recording and it's still doubtful there were any barnyard animals at Green Farm. Two nights earlier Burroughs had said he heard similar loud noises near the UFO landing site which sounded like women screaming. Could such sound effects have been part of something like a *son et lumiere* show put on to test reactions of the men out there?

The 18 minutes of Colonel Halt's tape recording which found its way into the public domain became a must-have item for UFO enthusiasts years later. It is certainly interesting, but there is also meant to be a huge amount of the recording he has never released. In 1999 he told author and investigative journalist Georgina Bruni that he had four or five hours of tape which nobody would ever be allowed to hear. When she pressed him very hard as regards what was on the tapes and why it could not be made available he refused to say any more.

Copies of the tape recording were stashed away by him, presumably as possible insurance should any official cover-up later denounce him as a liar or even seek to have him court-martialed.

This does make it seem as if he and his team experienced something even stranger than what he has told us. Maybe there was some extraordinary display by the supposed UFOs which utterly terrified the men? Not knowing what these things were or who was in control of a seemingly otherworldly display might have made the men think they were in mortal danger. And, since there was no explanation for what was going on, it could have made Halt think this really was the work of aliens—just like that UFO display at the end of the popular 1977 movie *Close Encounters of the Third Kind*!

Halt and his men were out in the forest for at least five hours or maybe six. All we have been allowed to hear about their time is the 18 minutes of recording on his tape. Is it just possible that Halt himself experienced something equally dramatic and extraordinary— perhaps similar to what Penniston and Burroughs experienced in the forest on the first night of the Rendlesham UFO saga? Or, could it be that what Halt is keeping secret on the tape, is a part of it during which he and his team became aware of unidentified men out there in the forest who appeared to be acting furtively? Although he would have known little of lasers and/or LPIs in 1980, could Halt have suspected these men, who may have come out from RAF Woodbridge, were possibly in control of, or else interacting with, the lights they saw in the sky?

Obviously that is only conjecture. Also, there's certainly no evidence that the craft referred to as the Rendlesham UFO seen in the forest on the first night descended there once more. Nevertheless, there are some people who seem anxious to prove that is what was happening.

One account in the book *Encounter in Rendlesham Forest* by Nick Pope, John Burroughs & Jim Penniston claims that Burroughs had a further encounter with the original UFO on the night of December 27th/28th. It is suggested a blue transparent light streaked towards John Burroughs and Adrian Bustinza and then "a white object kind of appeared up above and then floated down and was sitting out there in the distance". Burroughs was said to have requested permission from Halt to approach it. He did so, and it was said to have come towards them. "Sgt. Bustinza was on my right. He went down to the ground. He saw me go into the light. He saw me disappear. He saw the light explode and I was gone for several minutes before I reappeared..."

This particular account of what Burroughs was supposed to have said on the night of December 27th/28th sounds very much like a dubious rehash of his experiences two nights earlier. In this version, Bustinza is meant to be with him instead of Penniston. Then, John Burroughs was said to have "gone into the light and disappeared."

I suggest this particular story should be taken with a large pinch of salt! Burroughs himself has been quoted as saying "I have no recall of it. I have no memory of what happened. The next thing I knew I was standing in the field and whatever it was, was gone... What just happened?"

The story of this alleged further encounter with the UFO appears to have come from Adrian Bustinza, who quotes what John Burroughs supposedly said at the time, though the latter has no recollection of it. It is widely suspected that it was Bustinza who supplied Larry Warren with some of the sensational fictionalized account of the Rendlesham UFO's landing which we will look at in a moment. In any case, the above account from Adrian Bustinza has little if any evidential value since Burroughs himself has no recollection of the alleged event.

Whether Halt witnessed any such event involving Burroughs on the night of December 27th/28th we cannot tell. In any case he has been strangely reluctant to acknowledge Burroughs' presence on this further night of supposed UFO activity despite evidence from his own tape recorder. Burroughs had at first been denied permission to join Halt's team despite a request which he sent over the radio. Later he and Sergeant Bustinza were apparently allowed to meet up with Halt near the original UFO landing site. Then we have this story of Burroughs supposedly being authorized to approach one of the "lights" (presumably one down on the ground?) to ascertain if it was the same thing he and Penniston had encountered on the first night.

Colonel Halt has never confirmed or denied this story of an event at the edge of the forest probably during the early hours of December 28th. There's no reason to doubt his description of the other events he says he witnessed that night, though there seem to be additional things he has not made public. His sincerity is not questioned and nor can there be much doubt that he was reporting what he experienced and doing what he believed to be his military duty. But what did Halt really believe these UFOs were and who, or what, did he really think controlled the things which he and his men had witnessed?

Even at the time only two explanations seemed viable. Either this really had been an extraordinary and unprecedented visitation

AND PROJECT HONEY BADGER

by some extraterrestrial craft—or, it may have been a secret exercise carried out by a small detachment of US Special Forces men to test secret non-lethal weapons on unsuspecting airmen of the USAF security police at the twin bases. No other solutions seem plausible. Without actually saying so, Col. Halt may have thought it unthinkable that his own side could have carried out such an operation.

He repeated his belief that this was an ET visitation 30 years later in 2010 after he had left the military in the following terms:

> I believe that the objects I saw at close quarters were extraterrestrial in origin and that the security services of both the United States and the United Kingdom have attempted—both then and now—to subvert the significance of what occurred at Rendlesham Forest and RAF Bentwaters by the use of well-practiced methods of disinformation.

No one who is familiar with the case would deny that a thorough cover-up of what really happened in Rendlesham Forest that week was instigated by the US military authorities—and very likely by UK government authorities too, if any of the latter had been briefed on it. Whichever solution to the mystery was the true one, it soon became evident that whatever really happened in the forest was a secret that could never be told. The reasons why will soon become apparent.

It seems that very few, if any, of the commanding officers at the twin bases had any forewarning of these events. Certainly not Halt who was never told before or after what this was all about. The most senior officer, Colonel Gordon Williams, who commanded the 81st TFW, instructed that Halt's report on the UFO incident, the now infamous Halt Memo, dated 13 January 1981, should be sent to the British liaison officer for the bases, RAF Squadron Leader Donald Moreland. Col. Williams's thinking here was that these things had happened in the forest outside the Woodbridge base and that land was under British rather than American jurisdiction. Therefore any responsibility for it could be avoided and, with luck, the British MoD would lose interest in the matter and do very little.

That was largely true but over the next two years word got out and the strange story and rumors of the Rendlesham events reached UFO researchers and enthusiasts in both the UK and the USA. From CAUS in America came a number of Freedom of Information (FOI) requests to the US authorities. The first of these produced no result and CAUS were politely told any records relating to this incident did not exist since there had been no investigation by the USAF. A further request rather remarkably produced a copy of Halt's memo which had been kept by someone in the British MoD.

The copy was forwarded to the officer responsible for USAF document management services who happened to be a friend of Colonel Halt. When told, Halt pleaded with his friend to burn it, saying his life would never be the same again. The friend replied that FOI required him to release the document. This indeed did change everything for the unfortunate Colonel Halt and the apparent official confirmation that a UFO had landed in Rendlesham Forest gave rise to huge public interest in the matter. Whether or not Colonel Halt continues to this day to truly believe that the Rendlesham Forest UFO(s) were of extraterrestrial origin, we can only speculate on.

IS THERE ANY REASON TO THINK THE UFO WAS OF ET ORIGIN?

Being skeptical of extraterrestrial explanations is certainly reasonable but let us avoid blinkered skepticism based solely on the prejudice that ET UFOs and aliens cannot—and do not—exist. The latter hard-line skepticism—like that shown by the late Phil Klass and others—is as much a strongly held conviction as some of the false ET beliefs held by the UFO community.

Skeptic Phil Klass was frequently right when he debunked various UFO claims and particular cases but it also led to silly explanations for UFOs such as swamp gas—or perhaps a burning load of fertilizer as we saw earlier. There are of course many UFO cases which may have perfectly conventional explanations that just happen to be unknown. It is the tales of alien (extraterrestrial) contact that UFO investigators really need to be skeptical about since these invariably seem to be fraudulent and not one of has ever been proved true. Despite that, some in the UFO community believe that alien visitation of this planet is real and they believe in it with a religious fervor.

The reason for thinking Rendlesham UFO was not an alien spaceship is that there is a far better explanation which more closely fits almost all the evidence provided by the primary witnesses. It also explains the extraordinary behavior of the AFOSI and/or CIA/NSA agents who interrogated the primary witnesses after the nocturnal events of early December 26th through December 27th/28th during which time UFOs were seen. It explains too what UK Prime Minister at the time, Margaret Thatcher, had meant when she much later told Georgina Bruni in confidence that "You can't tell the people".

However, when the Halt Memo entered the public domain following the May 1983 FOI request by CAUS there was absolutely no stopping the story of the Rendlesham Forest Incident. Some of the British UFO researchers who had been chasing the story sold it and, of course, the Halt Memo to the UK Sunday tabloid *The News of the World*. That resulted in the sensational headline in its October 2nd 1983 issue which read "UFO LANDS IN SUFFOLK—AND THAT'S OFFICIAL".

The USAF authorities were, it would seem, blindsided by appearance of *The News of the World* story. Those who were aware of what had gone on nearly three years previously were horrified that the Halt Memo had got out. One can assume that much of their anger was directed against the unfortunate Halt. But the authorities were hardly in a position to deny what was stated in his memo and call him crazy or untruthful. Had they openly stated that, they would have had to explain what the RFI really had been about, and that was a secret that many in the military knew could *never* be disclosed. So they weren't exactly in any position to say that it was *not* of extraterrestrial origin. As with many awkward questions regarding state security during that era, the official line from both British and American governments would have to be "Neither confirm nor deny".

Rumors of the story of the Rendlesham UFO first became public knowledge within a few days of the events of December 26th– 28th 1980. Stories, some wildly exaggerated, began to spread among the local populace. Predictably the term UFO was interpreted as being something of extraterrestrial origin during the 1980s and the British public showed quite an appetite for such stories. Among the first to come out claiming he was a primary witness to the UFO landing was a certain Larry Warren (who initially used the pseudonym Art Wallace).

Warren was a young US airman who had recently completed his training and had been sent to the USAF twin bases in the UK. He claimed that he was one of about 40 men who were sent out into the field where the UFO landed and who witnessed small indistinct alien entities. These aliens appeared to be floating inside a translucent pyramid-shaped spacecraft. Its night-time arrival in the farmer's field (next to Green Farm) had evidently been expected since Warren claims to have heard over the radio the words "Here it comes—here it comes!" He also claimed that the Base Commander—presumably meaning either Gordon Williams or Charles Halt—then arrived and that he communicated with the ET aliens.

We were told all sorts of luminous UFO activity was seen in the sky, again sounding much like the scenes from *Close Encounters of the Third Kind*. Larry Warren also apparently suggested that this alien spacecraft had been damaged and that senior USAF personnel had agreed to help with repairs, presumably using avionic components from the A-10 airplanes which were stationed at the twin bases! We can say now without a shadow of doubt that Warren's story is entirely false. Particularly telling is that he was never mentioned as being present by any of the other witnesses who were sent out to look for the UFO(s). Although he was based at RAF Bentwaters at the time it is most unlikely he saw anything of the UFO(s) and his claim to have been a witness seems to have been based on what he heard from other US airmen. One suggestion is he heard stories of what allegedly happened in Rendlesham Forest from Adrian Bustinza—one of the men who were out there watching the sky on the night of December 27th/28th. These UFO stories had often been highly embellished.

In 1997 Larry Warren and Peter Robbins published a book about the Rendlesham Forest UFO called *Left at East Gate*. That same year I heard Warren speak at a UFO conference in Pensacola Beach, FL, and I knew instinctively that he was lying. Now, years later, most UFO researchers accept that Warren has been completely discredited and in recent years his foremost supporter, Peter Robbins, broke off all contact with him, realizing that much of Warren's story was untrue.

Nevertheless there are still many others who say they believe the Rendlesham Forest UFO case was indeed a case of extraterrestrial visitation, even if Warren's account of what happened there was mostly fictional. Foremost among them is, of course, Charles Halt.

Before we examine these things it should be repeated that there is a carefully researched book by Nick Pope, John Burroughs and Jim Penniston called *Encounter in Rendlesham Forest* that was published in 2014. This recommended book was written mainly by Nick Pope but it contains extensive contributions from John Burroughs, USAF (Ret.), and Jim Penniston, USAF (Ret.), who were of course the two primary witnesses in the case. The book is described as "the Inside Story of the World's Best-Documented UFO Incident". Although it examines every aspect of the case, including of course the possibility that the UFO was really of ET origin, it does not come to any firm conclusion as to what the Rendlesham Forest UFO, or UAP, was.

Nick Pope avoids giving any credence to the kind of alien spaceship type of UFO espoused by many traditional UFO celebrities such as Stanton Friedman, Linda Howe, or the late Budd Hopkins. He talks more of mysterious UAP (Unidentified Aerial Phenomena) that are associated with "plasma related fields" that can interact with human consciousness to produce visions of, say, huge triangular craft, etc. This is a new angle which seems to replace one UFO mystery with .... another mystery, the UAP. Whether extraterrestrial or interdimensional is again an unknown but the book definitely implies the Rendlesham UFO was something controlled by a non-human intelligence.

There is also the suggestion that senior US Air Force commanders are well aware of the "UAP" phenomenon and have developed ways of handling such situations when they occur. It should be pointed out here that neither Burroughs nor Penniston claim to believe it was anything like the popular concept of a flying saucer which they encountered in the forest. Penniston, at any rate, has suggested the UFO may have been a time-travel machine from Earth's own future!

WAS THIS "UFO" THE TEST OF A TROJAN HORSE WEAPON?

There is no compelling reason to think that the UFO which landed in the forest and the other objects seen in the sky at the time were of extraterrestrial origin. Apart from any other consideration, did some extraterrestrial UFO just *happen* to land right there by chance next to one of the principal nuclear-armed NATO bases in Europe ? If we do decide that it was not an ET spaceship or an ET drone then what are possible alternatives since this was clearly no natural phenomenon?

One proposal was that the UFO was a time-travelling craft that came back from mankind's future. It's a scenario which may have been suggested during regression therapy to one of the principal witnesses and which may have been accepted by him rather than any ET explanation. In my opinion it is extremely unlikely since there is absolutely no proof that time travel into the past is a possibility—or that any case of it has ever occurred. Apart from that, such a thing is logically inconsistent since it would allow time travelers to interfere with the past and thereby alter the future from which they supposedly came. The suggestion of time travel can be safely discarded.

So the only possibility that is left must be that the UFO which landed in Rendlesham Forest was something of human origin and—almost certainly—a craft with some military purpose in view of the fact it appeared near the perimeter of RAF Bentwaters/ RAF Woodbridge in December 1980. Since the twin bases were at the forefront of the military confrontation between NATO forces lead by the United States and the Soviet Union's Warsaw Pact forces, the likelihood is that it belonged to one or other of these two superpowers.

Could the UFO have been some exploratory craft used by the Soviets to probe the defenses of the Bentwaters/Woodbridge NATO base? That does seem unlikely as any probing on the ground could always have been tried by Soviet sympathizers or CND supporters as was the case at the RAF Greenham Common cruise missile base during the early 1980s. For the Russians to actually land a military drone or some kind of aircraft at the perimeter of a NATO nuclear base would have been foolhardy in the extreme and could have triggered a nuclear war. I, for one, believe that we can definitely rule out any possibility the Rendlesham Forest UFO was a craft of Soviet origin.

The alternative must be that the UFO which landed in the forest, and maybe also the lights seen in the sky that same week, were secret drones or other devices of US military origin being tested by a US Special Ops unit at RAF Woodbridge. If so, the UFO must have been deliberately placed there to attract the attention of the

USAF security police who were guarding the base. I believe the purpose of these non-lethal weapons was neither destruction nor reconnaissance but something of a specialized nature that was almost certainly planned for use on a top priority rescue mission far away from the UK.

I suggest the Rendlesham UFO that was approached by Burroughs and Penniston was what I call a THW—an acronym for "Trojan Horse Weapon". Its flashing lights were presumably meant to attract the attention of the base security guards on duty. Special Ops men, who would have operated such a THW, must have known that men of the 81TFW SP would see the lights and would know it was their duty to go into the forest to investigate whatever potential threat there was.

Also, the THW probably carried some kind of non-lethal psychotronic weapon intended to discombobulate or disable any person who approached it. That was achieved by zapping whoever came towards it with a directed burst of electromagnetic energy. The effect was to temporarily stun or subdue them physically and/or psychologically.

If the UFO was indeed a THW, it would have been deployed and operated by some elite unit of US Special Forces such as Task Force 160. They could have operated it just outside the Woodbridge base perimeter without the prior knowledge of anyone there—save possibly for Wing Commander Colonel Gordon Williams, the most senior officer then in command of the twin bases. I will return to the reasons for thinking the Rendlesham UFO was a THW later on but first let's look at what geopolitical or military situation in 1979/1980 could have concerned the US government quite so urgently. Was there was a requirement for just such a THW—if indeed that was what the landed UFO in the forest really was?

PART II

WHAT GEOPOLITICAL SITUATION IN 1980 MIGHT REQUIRE THE US MILITARY TO CONSIDER THE USE OF THWS?

During the 1970s the Soviet leader Leonid Brezhnev expanded the Russian military and the global influence of the Soviet Union grew dramatically. Cold War tensions between the Western alliance's NATO forces and those of the Soviet Union's Warsaw Pact forces rose to alarming levels which were perhaps only contained by the nuclear deterrent weapons which both sides possessed. The worldwide ambition of Soviet Communism caused some in the West to believe that a nuclear first strike by the Soviets was a distinct possibility.

To deter such a strike or, equally, to resist a massive tank advance across Western Europe by Warsaw Pact forces NATO built up both its conventional forces and its nuclear capability which was largely based in England and in Western Germany. Among its more important bases were the "twin bases" of RAF Bentwaters and RAF Woodbridge in Suffolk, England, which were entirely operated and controlled by the United States Air Force. During the years 1979-1980 the USAF's 81st Tactical Fighter Wing (81TFW) operated from the twin bases and the aircraft then being flown were mainly A-10 anti-tank aircraft or "tankbusters". These airplanes were designed to give close support to US ground forces, rather than being bombers as such, but the A-10s were certainly capable of carrying tactical nuclear weapons.

From 1975 onwards the Soviet deployment of SS-20 ballistic missiles in Eastern Europe caused major concern in the NATO alliance. The longer range, greater accuracy, mobility, and striking power of these missiles was perceived to alter the balance of security in Western Europe. It was feared the Soviets could launch a nuclear strike against Western Europe with a much reduced threat of any nuclear retaliation. NATO then devised a new strategy against this threat which involved both seeking arms control limitation talks with the USSR but at the same time deploying large numbers of ground-launched cruise missiles as a deterrent starting in the early 1980s.

In May 1979 Margaret Thatcher, Leader of the Conservative Party, won the UK general election and became Prime Minister. She was in office at the time of the Rendlesham Forest Incident in December 1980 and right through the 1980s. During these years she was US President Ronald Reagan's closest ally. Her support for the USAF military presence in Britain and allowing installation of nuclear cruise missiles at RAF Greenham Common and RAF Molesworth would shortly became a priority for her Conservative government but this was not without political opposition.

During the 1980s many socialists and Soviet sympathizers held demonstrations and marches to protest against Britain's nuclear weapons and against American nuclear weapons such as the cruise missiles that were being installed at these two US bases in the UK. The Campaign for Nuclear Disarmament (CND) claimed to be a non-political organization that was opposed to all nuclear weapons but there is little doubt it received much of its support from communist sympathizers since it never held any protests against Soviet nuclear weapons. Due to political hostility directed against NATO's USAF bases in England, the maintenance of a high level of security by the guards at such bases was certainly a matter of great importance.

This heightened tension of the Cold War during 1979-1980 and the resulting need for increased security at American bases where nuclear weapons were stored was certainly relevant. It was this situation at RAF Bentwaters/RAF Woodbridge in Suffolk that led some people to believe that the Rendlesham Forest UFO might have been a Soviet drone or missile sent there in order to reconnoiter the twin bases or to probe their security. However, that explanation is demonstrably untrue and it has since become clear the "UFO" was of US origin.

If the RFI was, as I suggest, the secret test of an American THW, it was carried out solely because the US government had an urgent need for just such a weapon in 1980. One reason for that was because of the dire US Hostage Crisis in Iran which had first gripped the United States in November 1979. Fifty two American diplomats and/or other US nationals were taken hostage in the US Embassy in Tehran and they were held captive for, eventually, 444 days. The siege continued through the whole of 1980, after a group of revolutionary Muslim students, followers of the fanatical Ayatollah Khomeini, first took over and ransacked the US Embassy. The militant college students, calling themselves the Muslim Student Followers of the Imam's Line, were soon backed by what became Khomeini's Islamic Revolutionary Guard Corps (IRGC) and the siege would turn out to be the longest recorded hostage crisis in history.

OPERATION EAGLE CLAW AND PROJECT HONEY BADGER

The take over of the embassy and subsequent imprisonment of the 52 American hostages led to an extremely grave international situation that consumed the attention of the United States, the President, and various arms of the US military for all of the year 1980. Diplomatic efforts to resolve the crisis failed and Khomeini's Revolutionary Shia Muslim regime demanded that the US send the ex-Shah,

Mohammad Reza Pahlavi, back to Iran to stand trial. That was something the US President refused to do. The Iran hostage crisis blighted Jimmy Carter's presidency and it concentrated minds both in government and in the US military to find some way of rescuing the American hostages alive. The use of elite forces in very specific operations against the enemy would become an alternative to costly wars and prolonged battle. Special Ops teams of highly trained and capable Special Forces could strike without warning, achieve their objectives and then rapidly withdraw with, hopefully, only minimal casualties.

Each branch of the US military, the Army, the Navy, the Marine Corps and the Air Force had its own Special Ops divisions. US Army Special Forces—the "Green Berets"—were originally known as Rangers and in 1977 a further operational detachment, Delta Force, was founded. 132 of these Delta Force and Ranger troops would be air-lifted into Iran and tasked with storming the US Embassy building in Tehran in order to free the 52 US hostages. The secret rescue plan was known as *Operation Eagle Claw* and was put into operation on April 24th 1980.

The complex operation would use eight large RH-53 helicopters and six large C-130 transport aircraft in this full scale attempt to rescue the hostages. The C-130s would fly the Army Special Forces soldiers to an improvised landing strip in the Iranian desert near Tabas, about 200 miles SE of Tehran. It had been secretly chosen and checked by CIA agents inside Iran in advance. It was codenamed 'Desert One'.

The plan was for the Special Forces troops to travel 260 miles on from Desert One in the RH-53 helicopters to a second staging point in the desert 50 miles from Tehran known as 'Desert Two'. After regrouping and keeping hidden there during the day, the men would rendezvous with CIA agents who were to bring trucks for their use. During the night the assault team would drive into Tehran before attacking the Embassy building. The plan was to eliminate the guards and IRGC men there and rescue the hostages with air support from AC-130 gunships flying from Desert One.

There was an equally complex plan for extracting the rescued hostages, and all of the Army Special Forces men, in helicopters from the Amjadieh Stadium in Tehran to, first of all, an abandoned airbase at Manzariyeh 60 miles SW of Tehran and then to fly them out of Iran to Egypt in C-141 Starlifter transport aircraft.

Operation Eagle Claw was to fail miserably with the loss of several aircraft and the lives of eight US servicemen. On the first night of Eagle Claw only five of the eight helicopters arrived at Desert One in an operational condition due to a desert sandstorm and mechanical problems. This type of sandstorm, known as a *haboob*, produced a vast almost opaque cloud of fine dust seriously reducing visibility.

The field commanders of both the ground and aviation forces decided reluctantly that the mission had to be aborted. Their recommendation was transmitted to back to the President in Washington by satellite radio. After two and a half hours on the ground at Desert One the presidential confirmation of abort was received. All the aircraft and their crews and the Special Forces troops were then instructed to fly out of Iran. It was at this point that one RH-53D helicopter moving on the ground collided, due to poor visibility, with an EC-130 that was carrying part of the Delta team. Both aircraft were engulfed in flames and eight servicemen in the two aircraft were killed in the explosion and ensuing inferno. After the crash it was decided to abandon the five remaining helicopters and their crews boarded the EC-130s which were now in the frantic process of departure.

Operation Eagle Claw was considered a defeat and a disaster. The leader of Iran's revolution, Ayatollah Khomeini stated the mission had been stopped by an act of God who had foiled it by sending his *haboob* to protect Islamic Iran and confound the American infidels. Subsequently, the 1979-1980 US Hostage Crisis in Tehran deepened still further. The embassy hostages were subsequently dispersed by their captors to several different imprisonment locations around the city of Tehran with the intention of preventing any further rescue attempts by US forces.

Nevertheless, shortly after the first rescue mission failed, planning for a second operation was authorized by the US government. The name for this second rescue mission was *Project Honey Badger* (PHB).

Plans and exercises were carried out but the sheer scale of such an operation grew to involve nearly a whole battalion of troops and more than fifty aircraft. Even though numerous rehearsal exercises were successful, the failure of the helicopters involved in *Operation Eagle Claw* resulted next in development of a different concept involving only fixed-wing STOL aircraft which were capable of flying from the US to Iran by the use of aerial refueling. After extracting the rescued hostages from Iran these STOL aircraft would return and land on a US aircraft carrier off the coast for medical treatment of the wounded.

This concept, known as *Operation Credible Sport,* was quickly developed but was never implemented. It was to use highly modified YMC-130H Hercules aircraft with rocket thrusters fore and aft to enable extremely short landing and take-off in the Amjadieh Stadium in Tehran. Initially the concept was to have been central to PHB.

The first of the modified YMC-130Hs crashed during a demonstration flight at Eglin AFB, FL, on October 29th 1980. That project was abandoned shortly afterwards and it was on November 2nd, 1980, the Iranian parliament set forth formal conditions for the US hostages' release—though few believed the Iranians would keep their word. At just this time Ronald Reagan was elected President of the United States to succeed Jimmy Carter, although, as is still customary, he would have to wait until eleven weeks later to take up office.

Because of the failure of *Operation Eagle Claw*, the US Navy saw the need for a specialized counter-terrorist team, USN commanders asked Navy SEAL Richard Marcinko to create an anti-terrorist force that could deal with this sort of crisis. SEAL Team Six was officially created in October 1980 and the team immediately embarked on an intensive training program. If there was to be a further attempt to rescue the hostages in Iran during 1981, SEAL Team Six could well have been the leading choice for such a dangerous operation. In more recent years SEAL Team Six has been seen as the nation's premier counter-terrorist unit—rather than the Army's Delta Force division. SEAL Team Six's reputation as the best of the best was vindicated on May 2nd 2011 by their successful execution of *Operation Neptune's Spear,* whose chief purpose was to kill Osama bin Laden.

Several years after the 1981 release of the US Tehran hostages, the existence of *Project Honey Badger,* the second projected rescue plan, was officially acknowledged. Even after the *Operation Credible Sport* concept was abandoned in late October 1980, alternative plans to rescue the hostages in early 1981 continued. It is not known which Special Forces teams would have been chosen to implement these plans or, indeed, what format would have been used. As we shall see, the plan for PHB was finally called off when President Reagan came into office and the hostages were at long last released by their Iranian captors on January 20th 1981, the day of his inauguration.

(For those unfamiliar with honey badgers it should be pointed out that this aggressive creature, said to be the most fearless animal on the planet, will kill and eat anything from fruit and honey to humans. The badger will attack and plunder honeybee nests oblivious to stings and it also seems impervious to the venom of snakes which it attacks. All of them appear to have no effect on its thick skin. Hence this very apt codename for a mission that was intended to go in and rescue the Tehran hostages. If the hostages were the honey and their captors were the hostile bees, any honey badger that rushed in would have to be utterly fearless and able to resist innumerable stings.)

It was eventually acknowledged that actual Honey Badger

exercises had continued for quite some time after the November 4th 1980 US presidential election. Several years later the USSOCOM (US Special Operations Command)—whose very existence was only officially disclosed in 1987 as having been founded and activated some years earlier—itself stated, regarding *Project Honey Badger*, that:

> *"Numerous special operations, applications, and techniques were developed which became part of the emerging USSOCOM (US Special Operations Command) repertoire"*

The new plans for US Special Forces in 1980 had also resulted in the formation of 160th Special Operations Aviation Regiment (Airborne), the elite *Task Force 160*, which became known as the 'Night Stalkers'. They, rather than the Delta Rangers, might well have been chosen to carry out a *Project Honey Badger* hostage rescue attempt operation.

Among the special operations and techniques being developed as part of *Project Honey Badger* one can assume there was a requirement for a THW that could be used immediately before any assault to rescue the hostages. Such devices could well have been deployed at night near the building(s) where the hostages were being held and their purpose would be to distract and deceive the Revolutionary Guards who would almost certainly come out to investigate. It is likely too that a THW intended to distract and confuse guards who approached it would also be able to neutralize them in some way. Then, a planned aerial assault by Special Forces units in helicopters would be met with much less resistance and might result in far fewer hostage casualties.

It seems most likely that the object which landed in Rendlesham Forest during the night of December 25th/26th 1980 was just one variety of THW that was being tested for PHB and that the USAF security guards on duty at the twin bases who responded that night unwittingly became the human guinea pigs for such a test.

Another possible variety of THW being tested at much the same time may have inadvertently lead to another infamous "UFO encounter" in a quite different part of the world: Texas. On the late evening of December 29th 1980 two women and the grandson of one of them were driving home when they encountered a "huge diamond-shaped flaming object" which hovered in the sky, then descended blocking the road in front. Both women subsequently suffered from what seemed to be heat radiation burns similar to severe sunburn.

This UFO was, according to these witnesses, accompanied in the sky by as many as 23 helicopters, including some CH-47 Chinooks. It was said that the object took off again and left the scene together with the accompanying fleet of helicopters. If it wasn't some secret test carried out by the US military which went awry, it is indeed hard to say what else it could have been. So let us look a bit more closely at what's called the Cash-Landrum Incident (**CLI**) and decide if that too could have been the testing of a THW being developed for *Project Honey Badger*. In view of the timing, any suggestion that this and the strange events in Rendlesham Forest during Christmas week 1980 were totally unrelated is, to me, very naïve to say the least.

THE CASH-LANDRUM UFO ENCOUNTER OF DECEMBER 29TH 1980

The story of what happened when Betty Cash, Vickie Landrum and her grandson Colby encountered a supposed UFO when driving at night in December 1980 near Dayton, Texas, is well known. However no satisfactory explanation has ever been found. Let us first briefly recap that event as related by Wikipedia:

> *On the evening of December 29, 1980, Betty Cash, Vickie Landrum and Colby Landrum (Vickie's seven-year-old grandson) were driving home to Dayton, Texas, in Cash's Oldsmobile Cutlass after dining out.*

At about 9:00 p.m., while driving on an isolated two-lane road in dense woods, the witnesses said they observed a light above some trees. They first thought the light was a plane approaching Houston Intercontinental Airport (IAH - about 35 miles away) and gave it little notice.

A few minutes later on the winding roads, the witnesses saw what they believed to be the same light as before, but it was now much closer and very bright. The light, they claimed, came from a huge diamond-shaped object, which hovered at about treetop level. The object's base was expelling flame and emitting significant heat.

Vickie Landrum told Cash to stop the car, fearing they would be burned if they approached any closer. However, Vickie's opinion of the object quickly changed: A born-again Christian she interpreted the object as a sign of the second coming of Jesus Christ telling her grandson, "That's Jesus. He will not hurt us."

Anxious, Cash considered turning the car around, but abandoned this idea because the road was too narrow and she presumed the car would get stuck on the dirt shoulders, which were soft from that evening's rains.

Cash and Landrum got out of the car to examine the object. Colby was terrified, however, and Vickie Landrum quickly returned to the car to comfort the frantic child. Cash remained outside the car, "mesmerized by the bizarre sight," as UFO researcher Jerome Clark subsequently wrote. He went on,

"The object, intensely bright and a dull metallic silver, was shaped like a huge upright diamond, about the size of the Dayton Water Tower with its top and bottom cut off so that they were flat rather than pointed. Small blue lights ringed the center, and periodically over the next few minutes flames shot out of the bottom, flaring outward, creating the effect of a large cone. Every time the fire dissipated, the UFO floated a few feet downwards toward the road. But when the flames blasted out again, the object rose about the same distance."

The witnesses said the heat was strong enough to make the car's metal body painful to the touch—Cash said she had to use her coat to protect her hand from being burnt when she finally re-entered the car. When she touched the car's dashboard, Vickie Landrum's hand pressed into the softened vinyl, leaving an imprint that was evident weeks later. Investigators cited this handprint as proof of the witnesses' account; however, no photograph of the alleged handprint exists.

The object then moved to a point higher in the sky. As it ascended over the treetops the witnesses claimed a group of helicopters approached the object and surrounded it in tight formation. Cash and Landrum say they counted 23 helicopters, and later identified some of them as tandem-rotor CH-47 Chinooks.

[These aircraft were used by the US Army and the US Marine Corps, besides other western forces worldwide. At the time these and other sorts of US military helicopters were known to be located at various bases in Texas. Betty Cash claimed that they could read "UNITED STATES AIR FORCE" imprinted on some of the helicopters.]

The story continues:

With the road now clear, Cash drove on, claiming to see glimpses of the object and the helicopters receding into the distance.

From first sighting of the object to its departure, the witnesses said the encounter lasted about 20 minutes. Based on descriptions given in John F. Schuessler's book "The Cash-Landrum UFO Incident", it appears that the observers were southbound on Texas state highway FM 1485/2100 when they claimed to have seen the object. The initial location of the reported object, based on the same descriptions, was just south of Inland Road, approximately at 30.0926°N, 95.1109°W.

Investigators later located a Dayton police officer, Detective Lamar Walker, and his wife who claimed to have seen 12 Chinook-type helicopters near the same area the Cash-Landrum event allegedly occurred and at roughly the same time. These other witnesses did not report seeing a large diamond-shaped object.

Above: Artist's impression of UFO seen by Betty Cash and Vickie & Colby Landrum on 12/29/1980

Soon after their close encounter with the mystery object Betty Cash and the two Landrums, once they had returned home, all suffered from nausea, vomiting, diarrhea and generalized weakness as though they had suffered severe sunburn. Betty Cash's symptoms were especially bad and worsened over the next few days. She could not walk and lost large patches of skin and clumps of hair causing her to be hospitalized. A radiologist was quoted as saying it appeared that all three patients could have been exposed to ionizing radiation (possibly from a nuclear source) or else to strong non-ionizing radiation such as microwaves or ultraviolet light.

Vickie Landrum telephoned a number of US government agencies and officials about the encounter. When she phoned NASA, she was steered towards NASA aerospace engineer John F Schuessler who had long been interested in UFOs and became years later the MUFON International Director. Schuessler researched the case in great detail and later wrote articles and the book about CLI. His undoubtedly sincere perspective of the incident was always that it must have been an extraterrestrial UFO or else, maybe, the testing by the US military of a recovered alien spacecraft like the alleged Project Snowbird.

After contacting their US Senators Betty Cash and Vickie Landrum were advised to file a complaint with the Judge Advocate Claims Office at Bergstrom AFB. They were next advised to hire lawyers and seek compensation for their injuries. Their attorney, Peter Gersten, who headed CAUS (Citizens Against UFO Secrecy), took the case *pro bono* and sued the US government for $20 million on their behalf.

On August 21ˢᵗ 1986, a District Court Judge dismissed their case noting that the plaintiffs had not proved the helicopters were associated with the US Government and that military officials had testified the United States Armed Forces did not have any large diamond-shaped aircraft in their possession. Colonel John B. Alexander was reported to have been one of the Army officers who gave evidence supporting the US Government's case that its military was not involved and was not responsible for what had taken place. Col. John Alexander is a friend of Lt. Col. George C. Sarran who gave decisive exculpatory evidence on behalf of the US Government in this case—to the effect that the US military did not possess helicopters like those which were described by Betty Cash and Vickie Landrum.

A DENIAL THAT THE US MILITARY WERE INVOLVED IN THE CLI

"There was no evidence presented that would indicate that Army, National Guard, or Army Reserve helicopters were involved."

That was the conclusion of Lt. Col. George C. Sarran's report on his investigation for the Department of the Army's Inspector General's office on the allegations of US military helicopters being present during the Cash-Landrum UFO encounter. His investigation was carried out 18 months after that extraordinary encounter on a road near Houston, Texas, and it is quite possible that he was unaware the above statement was untrue.

By specifying only "Army, National Guard, or Army Reserve" he may have been technically correct with this equivocation if the helicopters had been temporarily assigned to, for instance, the 160th SOAR(A) Night Stalkers. Despite what Sarran said, the actual truth about the presence of US military helicopters during the Cash-Landrum incident does seem to have been covered up or denied at all levels.

There is every reason to think that the helicopters that were seen by Betty Cash and the two Landrums were indeed US military ones and at the time they were engaged in a secret exercise of high priority. Without such secrecy and the subsequent denial given above, the whole purpose of the exercise might have been lost. Only the helicopter crews and the senior officers who ordered the exercise could have had any idea of its intended purpose. I suggest it was indeed a US experimental weapon being tested for possible use in *Project Honey Badger* which led to the extraordinary CLI encounter and serious health consequences for both Betty Cash and Vickie Landrum which may have led to their premature deaths.

If that were the case, we need to explain what the flaming diamond-shaped object was that descended on the country road near Dayton, TX. As with the UFO(s) seen by the USAF security guards from RAF Bentwaters and RAF Woodbridge, it seems very likely it was some kind of THW devised specifically for a further hostage rescue attempt.

Read that description of the CLI UFO again, and in particular this:

> *Every time the fire dissipated, the UFO floated a few feet downwards toward the road. But when the flames blasted out again, the object rose about the same distance.*

Does that sound like anything familiar? To me it sounds like a large hot-air balloon that rose when its burners, probably fueled by liquid propane, were turned full on and sank back down several feet when the gas burners were turned low. The inflated envelope of such a hot-air balloon above the burners could well be 70 ft high or so. It could easily appear to have the shape of a huge diamond in the sky. A regular hot-air balloon's burner(s) direct heated air upward inside its envelope. It, by night, would be illuminated by the burner(s). If the CLI UFO was indeed an unmanned hot-air balloon, it must have had additional downward-pointing burners since flames were seen shooting from its base. Considerable heat from such burners at, say, treetop level above them may have been the primary cause of the women's reported "severe sunburn". I reject any idea it was caused by nuclear radiation or by some DEW attached to the burner assembly.

Some may think the idea that the CLI UFO was a hot-air balloon like this, with a terrifying appearance, and flown at night accompanied by multiple military helicopters outrageous. Well, maybe rather less outrageous than the idea it was a nuclear powered alien spaceship being secretly flown and escorted by the US military who were likely in league with the aliens! The alien UFO explanation was in fact the thinking of some MUFON investigators who first researched CLI.

If this UFO was in fact a THW that was being tested for *Project Honey Badger* and its possible use in a further Iran hostages rescue attempt to be mounted in January 1981, the CLI encounter becomes a lot more understandable. An unmanned THW like this would have to be flown at night close to where the US hostages were being held to completely distract those guarding them and/or the IRGC men. The guards, if they weren't frightened out of their wits, would probably chase after the fire-belching balloon and rack it with gunfire. Bullet holes in a hot-air balloon's envelope would have little effect on its flight and would not send it crashing to the ground in the short term. However, the great alarm and confusion it would cause would permit US Special Forces in helicopters to approach from the opposite direction and then, hopefully, free the hostages who could be carried off to safety in, say, two of the large tandem-rotor CH-47 Chinooks.

Although the bold idea of using an unmanned hot-air balloon as a THW in this way may have seemed promising, execution of such a plan was obviously full of hazards. The rolled-up and uninflated balloon would have to be flown into Iran in one of the large CH-47 helicopters to a pre-selected spot near Tehran where it could be deployed. That would obviously have to be done on the ground. When sufficiently inflated it could be launched and sent on its way. The balloon's gas burners would, presumably, be under remote radio-control from a US helicopter crew. They would be able, in theory, to command the hot-air balloon to gain altitude or descend as required once it was flying in the intended direction.

However, the direction of flight usually depends on whatever wind there is and, if that wasn't properly determined the balloon might drift away in the wrong direction rendering the operation worthless. Perhaps the helicopter pilots had figured out how to make it fly in roughly the direction they wanted but the test near Dayton, TX, on December 29th 1980 seems to have gone badly adrift. That particular hot-air balloon may have descended, perhaps unintentionally, on to the road along which the unfortunate Betty Cash and the two Landrums just happened to be traveling at the time.

We don't know quite how close their car came to the balloon but all the balloon's downward-pointing burners were probably going full blast at one time for it to produce the heat effects they suffered which were likened to severe sunburn. Its regular upward-pointing gas burners would have been turned full on too in order for it to rise up again above the road—which it eventually seems to have succeeded in doing. There is absolutely no reason to believe that nuclear radiation or radioactivity of any sort was involved. The sort of high heat the two adults experienced would have been like standing too close to a hot bonfire or a furnace and could well have produced similar effects. In addition though, subsequent investigators have suggested the two women may have greatly exaggerated the skin burns they suffered in order to better justify their court claim for $20 million. It has also been suggested that Betty Cash had pre-existing ailments which could account for the health problems which she blamed on the encounter.

Although the CLI encounter was very traumatic for Betty Cash and the two Landrums there was no way the US military was going to admit to ownership of this THW. To have done so, its purpose would have been lost and the technique could never in future be used in, say, a hostage situation. Not that it's of any consolation to the victims, but I suggest any use of hot-air balloons as THWs was probably dropped after the CLI since it was evidently too difficult to control them.

Years after the mysterious Cash-Landrum Incident in Texas, I did once ask Col. John B. Alexander how he could possibly maintain that no US military helicopters had been involved in the encounter. And, unless Betty Cash and Vicky Landrum had made the whole story up—which no one believed—what were the military helicopters they saw?

He answered that it seemed entirely possible to him that what the women saw may have been some kind of moving VR (virtual reality) scenario that was constructed for them by the aliens in order to mask this UFO from human observation.

I, for one, don't buy such a fantastical explanation and I really do think even the most devout ET UFO believer would probably have some difficulty in swallowing that!

COULD SUCH THWS BE USED TO ASSIST HOSTAGE RESCUE OPERATIONS?

Any modern THW would have to be a totally unfamiliar object and one whose purpose was not obvious to the people it was intended to fool. Its objective would be to deceive, distract, and probably even disable enemy defenders at the position under attack. I don't suggest that it would necessarily be meant to look like a UFO but any such THW would very likely have to descend from the sky at night, land close to where a hostage rescue operation was to take place, and distract—even neutralize—those who were guarding the hostages. If this was to work in 1980/1981, the hostages would be rescued from the large building where they were being held in Tehran. Task Force 160 men would descend onto the roof of the building from helicopters and blast their way into the place where the hostages were being held.

Although he does not use the term THW, Jacques Vallee has explored this concept in books he has written such as *Messengers of Deception* (1979) and *Revelations—Alien Contact and Human Deception* (1991). Among the cases where such deception may have been used, he asks how a small military intelligence unit could simulate complex UFO events and cites the Rendlesham case (which he refers to as Bentwaters). Secondly, Vallee asks "Why would they want to do it?" From his book *Revelations*, he has already told us:

The devices in question can be equipped with mechanical, optical, and electronic devices that can be used in sequence or in combination to produce very spectacular UFO sightings.

The simplest such device is a model of a disk, two to four feet in diameter. We are not talking here about crude garbage-can covers equipped with hobby rockets, but exquisitely controlled systems carrying microprocessors and guided by radio. Miniature television cameras enable these gadgets to survey their surroundings and to transmit pictures. They can maneuver in and out of trees . . .

Of course we are all familiar with today's drones which come in many different shapes and sizes and are controlled by micro-computers that are either preprogrammed or remotely controlled by some operator at a distance. The word drone was little used back in 1980 but much of the technology was available then, as Vallee says. He then poses the question with regard to UFO events like those in Rendlesham Forest: Why would a small military intelligence unit want to do that?

As he suggested, a number of purpose built THWs including ones looking like UFOs that remained airborne could have been of value in a hostage rescue situation like that in Iran in 1980. However, I doubt that actual drones as proposed by Jacques Vallee were the UFOs that were seen by Colonel Halt and his men. Quite apart from physical UFO-like drones, there is no doubt the US military during the Cold War years developed a repertoire of laser effects—projecting lights or images into the sky in order to distract or confuse the enemy.

If the Rendlesham UFO and the Cash-Landrum object were indeed different sorts of THW then unsuspecting witnesses may perhaps, unsurprisingly, have thought these were UFOs in the alien spacecraft sense of that acronym. That may well have been anticipated by the US military operators of any such THWs and they may have been quietly pleased that these alien explanations diverted attention from the true nature of their extraordinary weapons.

HOW SPECIAL OPS MIGHT TEST THW BEFORE A SECOND
RESCUE ATTEMPT

In order to test a THW on unsuspecting guards with a role
perhaps somewhat similar to that of the Revolutionary Guards
in Iran, one possible option would be to try it out on the security
guards at some US military base. Such an exercise would allow the
thorough interrogation of the guards involved by US intelligence
officers so the THW's effectiveness, or otherwise, could be quickly
assessed.

One might ask why such a THW test couldn't be carried out at a
US base in America rather than at RAF Woodbridge in the UK. The
main reason for that was because the 67th ARRS with its specialist
HH-53 helicopters was based at Woodbridge. It was that particular
squadron which apparently was to be tasked with infiltrating the
THW into Iran if a planned new rescue attempt was to go ahead.
Their pilots already had experience of flying with BP CMs slung
below HH-53s. An additional bonus to testing a THW in England,
rather than in the USA, was that the 81TFW security guards were
supposed to leave their firearms inside the perimeter of the USAF
base if they ever needed to go outside the base. In the US unwitting
base security guards would probably have little compunction about
opening fire on an unfamiliar craft—or indeed any unidentified
person thought to be associated with it—if it were to land in or
close to a military base.

The actual hostage rescue operation could be rehearsed
separately—perhaps by a unit of the 160th SOAR(A)—and
that would be done in the United States. The Eagle Claw rescue
operation had previously been rehearsed in the US and a further
rescue attempt would also have been. Likewise, practice testing of
the CLI object, if that was indeed a Honey Badger THW for use in
Iran, could be done at night in Texas.

The choice of the Bentwaters/Woodbridge USAF base would also be logical since it was a closely guarded base on foreign soil thousands of miles from the United States and nobody would have any suspicion of what this secret exercise was. In fact it is most unlikely that anyone at the twin bases, apart from the 67th ARRS commander Colonel Wicker and his helicopter pilots who were involved, had any idea of what was going to happen in advance. Presumably the senior officer at the twin bases, commander of the 81st TFW at Bentwaters, Gordon Williams, would have been told there was to be a secret exercise outside the perimeter of RAF Woodbridge, but also would have been told to say nothing about it until he received further orders.

I suggest that the Rendlesham Forest venue—just a mile outside the perimeter of the Woodbridge base—was carefully chosen as the best spot to place the THW. It obviously had to attract the notice of the base security guards and it must have been anticipated that a small number of guards would come out from the Woodbridge base to investigate. The action would obviously have to be in the middle of the night so that only the target audience—some of the base security guards—would see the unexplained lights and only they would go out to check on what these were. The position where the THW was placed in the forest ensured it could not be seen from any public road.

The timing of such an exercise was no doubt carefully chosen too. During Christmas Week 1980, starting on December 25, there was no military flying activity scheduled from either of the twin bases and the majority of USAF personnel would be off duty or away. No aircraft movements would interfere with the planting of a THW in the forest or its swift removal following the test.

As we have seen from the many accounts of the Rendlesham UFO visitation from December 26 through December 28, the secret exercise by a small detachment of Special Forces personnel seems to have continued on the following two nights. Although the actual

THW itself didn't make a further appearance, it seems that these men used lasers to project moving lights in the forest and in the sky to confuse and misdirect the airmen who were sent out in the forest to search.

That does seem a much more likely explanation of what occurred than the ET one of an alien spacecraft landing in that particular place quite at random in the early hours of December 26th. An even more absurd explanation that was suggested at the time was that ET aliens had suffered some kind of a flying saucer breakdown and landed in the forest to borrow some A-10 avionics from the friendly USAF!

Further confirmation that it was indeed a secret exercise authorized by the highest levels of the US military can be inferred from what Colonel Halt has clearly stated about an unannounced visit to RAF Woodbridge soon after the events of December 26th - 29th 1980. A large C-141 military transport aircraft arrived there but great secrecy surrounded the purpose of its visit. Halt was completely unable to ascertain who was on board and what their mission was. Apparently a group of "special individuals" disembarked, passed through East Gate, and headed out into the forest where the UFO encounter had taken place a few days earlier. In retrospect, "special individuals" most probably meant senior officers of the US Special Forces.

It was said that General Charles A. Gabriel—the CINCUSAFE, no less!—made such a visit to the twin bases at this time and it was very probably connected with the Rendlesham Forest Incident. It seems Colonel Williams and Colonel Halt were not even briefed on the visit since they were evidently not in the loop as regards the planning for *Project Honey Badger*. Since their visit to Woodbridge was in a C-141 transporter aircraft, the trip could also have been used to collect the Apollo boilerplate CM (and its specialized DEW device) that had been perceived as a landed UFO in the forest clearing. This hardware could then have been flown back to the USAF's Ramstein AFB in Germany (where CINCUSAFE was based) with a view to its possible immediate deployment in a new attempt to rescue the US hostages in Tehran. That never happened but, if it had gone ahead then, the same Special Forces unit might have flown on to Iran to operate this THW.

In fact, it probably never left Woodbridge for Ramstein. When Honey Badger was discontinued and it was no longer required, it must have been stripped of any special equipment such as a DEW and returned to the custody of 67th ARRS at the base. From 1981, and for the rest of that decade, the official line was it had never left RAF Woodbridge—not even in 1977 when this BP CM had been flown back to Florida.

MYSTERIOUS LIGHTS OVER KIRTLAND AFB IN 1979 & 1980

Quite apart from THWs which I suggest played a central role in both the Rendlesham Forest UFO mystery and the Cash-Landrum Incident at the end of December 1980, it's worth looking at what was going on at Kirtland AFB, NM, during the years 1979/1980. Though digressing somewhat from the main theme of this book we find multiple cases of supposed night-time UFO sightings were reported at a place where several top secret military projects were in progress. At the time few outsiders had any idea that these projects involved the development of lasers for military use. From outside the base's perimeter it did look as if UFOs were flying in and out, maneuvering, and sometimes appearing to land at the huge base. These were certainly not aircraft.

Close to the gates of Kirtland Air Force Base in Albuquerque, New Mexico, was the home of scientist Paul Bennewitz. There too were the premises of his company, Thunder Scientific, manufacturers of specialized humidity and temperature measuring instruments for high-profile clients such as NASA and the US Air Force. As CEO of this successful outfit his dealings with the military were made easier by the fact they were also his close neighbors.

On a cold winter night in 1979 Bennewitz stepped out on his deck for what was becoming an almost nightly ritual. Since September he and his wife had been seeing multicolored lights floating in the sky or else shooting across the small range of hills inside the secure base area about a mile away. Through massive telephoto lenses Bennewitz shot hundreds of photos of these mysterious luminous objects and took thousands of feet of film. Often the lights would streak away at impossible speeds only to re-appear seconds later seemingly miles from where they had just been. Over several weeks he also recorded lengthy electronic signals that seemed to be coming from these mysterious UFOs and/or transmitters inside the base.

Paul Bennewitz became increasingly convinced that these night-time UFOs were landing and taking off in the most secretive and highly-guarded area of Kirtland AFB known as the Manzano Weapons Storage Complex. This was, at the time, the largest underground repository of nuclear weapon components in the Western World. To him it was clear evidence that something unearthly was playing cat and mouse with the US military and daring it to react. If this was the first stage of an alien invasion Bennewitz felt he was duty-bound to alert the powers that be.

Through his contacts at Kirtland AFB Bennewitz offered to present his huge tranche of photographs and electronic recordings to senior USAF officers to alert them to what was happening and to the possibility that these UFOs suggested some kind of alien invasion was underway. Although Air Force officers appeared dismissive at first, they invited him to give a presentation at the base a few weeks later.

On November 10th 1980 Paul Bennewitz drove into the grounds of Kirtland AFB with his photos, films, and tape recordings to make the presentation to the top brass at the base. In the meeting room he was introduced to base commander Brig. Gen William Brooksher, four colonels, Thomas Cseh (Head of AFOSI at Kirtland) and another AFOSI agent there. There were also two scientists from the Air Force Weapons Lab which is headquartered at Kirtland. For most of the presentation the audience had shown interest in the photos and the electronic signal recordings but when he began to mention UFOs, aliens, and his suspicions of an alien invasion, some left the meeting indicating they had pressing business elsewhere.

Paul Bennewitz's scientific credentials were not in doubt but it seems evident that some of the Air Force officers must have thought him a little crazy. Nevertheless they decided that AFOSI should continue what was described as a scientific investigation of Bennewitz and analyze the signals which he was picking up in case he had been eavesdropping on secret Air Force projects. It was significant that the most interested parties were associated with NSA operations on base.

Apart from what Bennewitz claimed was happening at Kirtland there were others who were seeing UFOs there too. From Greg Bishop's book *Project Beta* we have the following report which is taken from the official 'Kirtland Documents' released under the Freedom of Information Act (FOIA) in 1983:

On September 2, 1980 at about 11.30 p.m. three Air Force security policemen saw a bright light swoop over head at high speed and stop suddenly over Coyote Canyon. It is difficult to convey the strangeness of this sort of movement, but it is often described by witnesses as appearing like the UFO is at the end of an invisible flashlight beam that is swept across a wide sweep of sky. The thing appeared to land and stayed on the ground sometime before leaving by shooting straight up . . .

[The documents refers to bunkers in the area as storage for "HQ CR 44" material. "Headquarters Collection Requirement 44" refers (or referred to) nuclear weapons and their security.]

If this is a genuine and accurate report from a Kirtland security guard from September 1980 one should ask whether this "bright light" UFO was in fact a false UFO of the kind I call a LPI. This was at a time when few knew of the capabilities of powerful lasers and most people spoke in terms of more familiar searchlight or flashlight beams. The fact that some laser beams are quite invisible when seen from the side would add to the illusion that the bright spot of light projected by such a laser was itself a real luminous object or actual physical UFO.

During the Cold War in the late 1970s there was much talk of particle beam weapons which some saw as a definite game-changer in any future conflict. In theory, an orbiting satellite equipped with such a device could fire beams of charged subatomic particles to destroy targets down below such as ships, airplanes, buildings and even towns. Conversely, a ground-based particle beam weapon could be fired into the sky to destroy aircraft, incoming ICBMs, and, possibly, enemy satellites which orbited over one's territory.

Major General George Keegan, a former Air Force intelligence chief in Washington, constantly warned the Soviets were ahead in developing such weapons and they apparently had a secret experimental facility located in a remote part of the USSR. Keegan had previously worked for many years at the Pentagon but most of his former colleagues publicly disputed his claim that the Soviets were ahead. Nevertheless urgent studies would be carried out into the feasibility of such weapons and much of the work was done at the Air Force Weapons Laboratory (later known as the Phillips Laboratory) at Kirtland AFB.

Although outsiders had little idea of what research went on there, we now know that the scientists soon realized the concept of an effective particle beam weapon was unrealistic to say the least. Huge amounts of power, maybe even controlled nuclear explosions, would be needed to send out a particle beam that would destroy target aircraft or missiles a hundred miles away or just tens of miles distant from such a supergun. Quite apart from that, the expense of producing such weapons would run into many billions of dollars.

However the scientists at Kirtland soon saw there were several promising applications for military laser weapons that were capable of interfering with, if not actually physically destroying aircraft or missiles, at ranges of up to a hundred miles or so. In 1980 several different government agencies were working on different laser projects at Kirtland. Sandia Labs and Phillips worked on some of the most sensitive and advanced defense projects in the US. The NSA was also involved in several different secret laser projects there as well.

The Air Force Weapons Laboratory pioneered the first and only megawatt class airborne laser at Kirtland. It also built the Starfire Optical Range (SOR) at Kirtland AFB. This is a USAF secure research laboratory to develop and demonstrate optical wavefront control technologies using adaptive optics. The facility was run for many years by Dr. Robert Q. Fugate who was a top scientist and a specialist in optical physics from Sandia Labs.

He had a particular interest in astronomy and had worked at the famous VLT (Very Large Telescope) situated high on a mountain in the Atacama Desert, Chile. The reason for placing VLT there was to overcome, as best as possible, the eternal problem with powerful telescopes that are based on earth rather than above the atmosphere like the Hubble Telescope which orbits in the vacuum of space. The constantly moving molecules of turbulent air cause images of distant stars and galaxies in telescopes to be distorted and blurred to a greater or lesser extent. The calm cool air above the VLT minimizes the problem but any degradation of images caused by poor viewing conditions renders, in general, much telescopic work useless.

Fugate, who many fellow scientists considered to be a genius, was one of the first to develop a computer-controlled adaptive optics system (AOS) to overcome this problem. In the late 1970s Fugate and his team at Kirtland built their AOS in the form of a very thin mirrored surface that could constantly be slightly deformed by an array of plungers attached to the underside of the surface. All the tiny plungers were instantaneously controlled by a powerful computer. The system resembled an orthodox reflecting telescope with a parabolic mirror but the adaptive optics constantly compensated hundreds of times a second for distortion of stellar images. Such an AOS could also be used to produce crystal clear images of Soviet satellites orbiting in the skies a hundred miles or more above New Mexico.

It was the latter use of the Starfire Optical Range at Kirtland that was so important to the US military and the intelligence agencies at the peak of the Cold War in the 1980s. The SOR was of course a closely guarded military secret and its very existence was only revealed to the public over 20 years later. Having developed an ability to closely monitor the purpose and behavior of Soviet satellites, the consequent intelligence enabled the USAF to explore whether laser beams from Kirtland could also be used to destroy or disable them. The success or otherwise of the tests which they carried out to do that is still the subject of speculation.

Another problem other than atmospheric distortion faced Dr Fugate's team trying to photograph Soviet satellites. In calibrating the AOS they were using, some fixed stars in the night sky were required as guide stars for the computer to have anchor points, since it constantly measured the degree of changing atmospheric disturbance. Only a few bright stars were really suitable for this purpose and the target satellites moved fairly rapidly across the sky in their orbits making the SOR adaptive optics less stable than was desired. The answer to the problem was a Laser Guide Star (LGS) and it was Dr Fugate who realized this had to be the cornerstone of adaptive optics. He and his team developed a method of shooting a bright laser beam into the sky alongside the telescope that was being aimed at a passing satellite. It produced a bright artificial star high up in whatever position was required. That could be moved across the sky in the same direction as the target satellite moved. By observing how the exact position of this LGS varied, precise measurement of atmospheric distortion could be made and applied instantly by the computer controlling the AOS.

Unlike regular astronomical telescopes, SOR was able to track and photograph enemy satellites during daylight by using suitable filters. This meant NSA—or maybe DIA—agents could photograph Soviet spy satellites at the same time as these satellites were photographing the base at Kirtland AFB with its various secret installations. SOR laser beams would generally have been invisible—both by day or by night—to people on the ground, like Bennewitz, but a LGS moving across the night sky might look like a bright satellite—or to some like a UFO. In addition, the powerful lasers at SOR were most likely being used to blind or even completely disable Soviet satellites.

Despite the cool reception of Bennewitz's presentation to Air Force top brass at Kirtland in November 1980, they and NSA agents there decided that he should be encouraged to continue with his electronic surveillance. From early 1981 Bennewitz was listened to and visited on a regular basis by Kirtland's AFOSI agents, in particular the notorious Richard Doty. The latter, while absolutely claiming that he believed Paul Bennewitz was seeing UFOs and recording their communications, fed disinformation on the "alien presence" to him. Consequently, other UFO investigators, such as Bill Moore and Linda Howe (who met Doty at Kirtland AFB in 1983) heard all about the supposed alien threat, if not quite an actual alien invasion!

Much of modern American UFO myth, such as the supposed ET UFO crash at Roswell in 1947, came from disinformation originating from the Kirtland AFOSI operation with which Doty was associated. There was also the US government's alleged secret MJ-12 committee that had supposedly been set up in 1952 to investigate ET aliens dead and alive which they supposedly referred to as "Extraterrestrial Biological Entities (EBEs)". These EBEs were said to have been recovered from the 1947 Roswell crash (or other flying saucer crashes). Doty falsely claimed there was a secret pact between the US government and the aliens. He alleged they had secret underground bases and that the killing and mutilation of cattle—besides the abduction of human beings—was being carried out by the extraterrestrials in their UFOs.

In 1981 when Bennewitz was writing to his New Mexico senators, Pete Domenici and Harrison Schmitt, about the alien invasion that he felt sure was going on, AFOSI Special Agent Richard Doty, often together with Bill Moore, frequently visited his house near the base sometimes letting himself in when Bennewitz was not at home. The array of electronic surveillance equipment that was focused on the base was itself photographed and recordings and photographs from Paul Bennewitz's home lab were examined and analyzed.

On one occasion it was said Dr Robert Fugate himself was actually smuggled into the house when Bennewitz was away. Excited at being involved in this cloak and dagger stuff, Fugate examined a device which Bennewitz had cobbled together to measure variations in magnetic fields in the vicinity of the UFOs he saw above the base. Since the Air Force and the different intelligence agencies involved with their own secret projects at the base didn't usually share data, none of them had a full picture of what was really going on inside the base. They each recognized some of the Bennewitz's recordings but suspected some unfamiliar signals might relate to Soviet espionage.

In December 1981 Bennewitz wrote at length to President Reagan about his fears and suspicions of an alien invasion. He urged the President to act but a full month later all he got back was a bland standard letter saying that previous Air Force investigations of UFOs had concluded that there was no threat to national security.

Evidently the Air Force and the intelligence agencies involved were both unwilling and unable to tell Bennewitz that they were the ones who were responsible for the strange lights and supposed UFOs that he was seeing over the base and the mysterious electronic signals that he was picking up. Had they done so he would never have believed them and it was already too late trying to stop him from putting out word of the supposed alien invasion to APRO, to MUFON and to the American UFO community at large. The authorities were also aware that Soviet spies in the US who knew that secret military testing was carried out at Kirtland might mix with these UFO enthusiasts who were taking an unhealthy interest in the base's activities.

Bennewitz was visited on several occasions by AFOSI agents other than Richard Doty who were most probably NSA or DIA men but preferred not to mention that was the case. An empty house across the street was unobtrusively taken over by one of these agencies, most likely NSA, and Bennewitz himself and his visitors were kept under surveillance for many weeks. He had put his total trust in the AFOSI at Kirtland and, in retrospect, it can be seen that they deceived him and played him at every turn from the word go.

A few months after Bennewitz's first meeting with the top brass at Kirtland, the Air Force and their AFOSI people tried to draw his attention away from the base. He was told that it was known alien activity now seemed to be concentrated in the Dulce, NM, area some 130 miles north of Kirtland. He was told there was believed to be an alien underground base below Archuleta Mesa in the hills above Dulce and that UFOs had been seen flying in and out of this place. Moreover, there had been several reports of mysteriously mutilated dead cattle found in this remote region which locals believed to be connected with the UFO activity. Doty even arranged for Bennewitz to be flown over the Dulce area and pointed out to him some of the supposed sites where alien UFOs had landed or flown into openings in the hillsides. There were several further helicopter flights with Bennewitz being flown directly from Kirtland to Archuleta Mesa.

Doty says the Air Force airlifted old storage tanks, engineless jeeps and other military equipment into this inaccessible rough country near Dulce to fool people like Bennewitz who flew over it that there was a hidden alien base down below. He says that Robert Fugate was asked to come up with some method of projecting flying saucer like shapes onto clouds in the Dulce area in order to fool Bennewitz and other UFO investigators who were drawn there. The projector he devised was said to consist of a rotating lens mounted on searchlights. If this is true, I feel we can be sure the "searchlights" described by Doty were probably lasers, which were of course Fugate's speciality. It may even be that, in 1980/1981, Robert Fugate was the father of what I am now calling the Laser Projected Illusion (LPI).

This extraordinary charade which was principally mounted by AFOSI at Kirtland AFB—and by Richard Doty in particular—continued to deceive Paul Bennewitz and other UFO investigators for months and years to come. The deception became the foundation for much of the UFO myth about the "alien presence" and "alien abductions" that was developing during the 1980s. There can be little doubt it had serious effects on Bennewitz's mental health leading almost to a nervous breakdown. Had he ever realized the falsity of the disinformation he was being fed by AFOSI, he might well have been able to sue the Air Force for substantial damages. Richard Doty claims that at a late stage in the 1980s he decided to come clean with Bennewitz and admit much of the stuff he'd previously told him was disinformation. "It was my job," said Doty. "I was ordered to lie to you". Bennewitz simply said in reply "No, it wasn't." Paul Bennewitz died in Albuquerque, NM, aged 75, in 2003.

LASERS COULD EXPLAIN MANY UFOS SIGHTINGS OF THE 1980S & 1990S

About twenty five years ago I purchased a laser pointer. It was a small hand-held device three inches long and powered by a single cell battery. Its intended purpose was use as a pointer when presenting lectures to audiences. From a range of perhaps 20 feet it produced a small red dot on a film projection screen or, say, a display board.

During the 1980s a variety of lasers were commercially available but most were usually larger, heavier and more expensive.

Besides using the pointer for lectures there was also a very different use for it that I discovered quite by chance. Staying one night with a friend in a treehouse in the rainforest of eastern Maui, we found that this little laser could project its bright red dot a good 200 feet or even further. The projected dot could be moved slowly across the ground below and sometimes made to climb some way up trees in the forest. One amusing trick with the laser was its effect on passers by who had absolutely no idea what they were seeing.

The beam of the laser pointer was completely invisible and anyone seeing it would simply see a moving red dot. Unwitting passers by would be mystified as to what the red luminosity was or where—if anywhere—it was coming from. They were unaware of our treehouse or even that there was anyone else in the forest late at night. There were few who walked on the paths through the rainforest at night but some who did appeared to be seriously scared. Maybe this luminosity was similar to the "strange small red light" which Colonel Halt said he saw among the trees in Rendlesham Forest in December 1980?

Although we never emerged from our treehouse to ask anyone who saw the moving red laser dot what they thought it was, it looked as if they perceived it as something akin to a mysterious nocturnal animal or, maybe, a tiny UFO. Like many UFO reports, the sighting of a mysterious red luminosity in the rainforest late at night may have always remained a mystery to those who saw it.

The same trick with the laser pointer was much less impressive when tried in an urban environment where there were more people about than in the forest. If the red spot was projected on, say, the wall of a house passers by at night might pause briefly to watch but usually shrugged and moved on assuming this was some kind of human activity. For the same reason it is fair to say that few impressive UFO sightings occur in towns or cities with multiple witnesses since most onlookers would make similar assumptions. Out in the forest or in the desert where there seems to be no one about, a moving light in the sky—if it's clearly not behaving like an airplane or a satellite—is sometimes taken as a UFO. Equally, it *could* be a laser light projected by someone unseen. Small inexpensive lasers are common enough today and one can even buy a pet toy laser pointer in Walmart to tease one's cat or dog for just $5. How unlike the 1980s when folk were far readier to accept that any strange light in the sky was a UFO!

The extraordinary UFO sightings by Colonel Halt and his men in Rendlesham Forest during the small hours of December 28th 1980 could certainly have been similar laser effects produced by just two or three Special Forces men who had concealed themselves in the forest. Small circular or oval luminosities moving in the sky at the time could well have been produced by laser beams. I don't say they necessarily were but that does seem far more likely than alien UFOs.

Critics of this argument may say that small laser pointers of this type were not available back in 1980 and so it isn't a plausible explanation. The first working laser was built and tested in the US by Theodore H. Maiman at the Hughes Research Laboratories California in 1960. During the next 20 years most lasers used for scientific research were known to have been fairly large devices and weren't widely available to the public. However, the US military were quick to see the many possibilities afforded them by lasers and much secret research was undertaken to produce weapons that used the new technology. Some of the applications were target designation and ranging (including gunsight lasers), and also directed-energy weapons (DEWs) which were intended to destroy or interfere with enemy cruise missiles and aircraft. Smaller portable lasers for use by soldiers in the field were certainly available by 1980. However, the development of particle beam weapons and "death ray" lasers to destroy enemy missiles or airplanes proved largely unsuccessful.

In 1980 most people, including many in the military, would have been unfamiliar with lasers of any kind and the bright spots of light that lasers can produce on material surfaces or in the sky above. Unlike the searchlights used for anti-aircraft defense in earlier eras some lasers have a completely invisible beam though with others it may be easily visible when seen from a position off to one side. Visibility of the beam from the side depends on whether the light from the laser is scattered—or not—possibly by fine dust particles, mist or water vapor, or haze in the atmosphere. It depends indeed on atmospheric conditions but also, primarily, on the wavelength of the laser light used. The human eye is much more sensitive to green or blue light so, if that is scattered sideways from the main beam, the beam itself may be more visible unlike a beam from a red or orange laser.

Obviously, a laser beam projected at low cloud or even thin layers of mist or haze can produce a small disk of light in the sky which could easily be mistaken for a UFO. Certainly, during the 1990s laser lights—often so-called disco lights—moving rapidly in the night sky near centers of population were a familiar enough sight in parts of the UK and other countries. These are seldom seen in the present day since there are now laws to prevent people shining bright lasers into the sky in case they blind airplane pilots and maybe cause a plane to crash.

Nevertheless, pointing a powerful laser into the night sky can sometimes produce a small luminous disk that looks like a bright star or, if the laser beam itself is not visible from the side, possibly like a UFO. Such disks of light can appear by reflection from an invisible layer of mist or haze especially where a temperature inversion has occurred at some level in the atmosphere. Atmospheric pollution above large cities sometimes contributes to layers of haze like this that may reflect back light from a laser.

As we have seen it's perfectly possible to produce an artificial star—looking like a UFO—by projecting a laser in a clear night sky. Such a luminosity high up in the night sky is the same as what astronomers term a laser guide star (LGS) that they create for use in adaptive optic systems (AOS) employed in large telescopes. A laser is projected into the night sky to produce such an artificial star image: light from the laser is reflected by components of the upper atmosphere back into the telescope. The LGS works because laser beams experience light scattering solely by air molecules at certain altitudes and that is one form of what's called Rayleigh scattering. It is chiefly from a sparse invisible layer of neutral sodium atoms which have accumulated from ablating meteors entering earth's atmosphere about 55 miles up. The band is thought to be mainly responsible for this type of Rayleigh scattering and allows production of such artificial laser guide stars.

It can occur too when one's line of vision is at a fairly close angle to the axis of a laser beam. That allows one to see light scattered back along much of the length of a narrow laser beam that would be far less visible than from well out to the side. It could mean a false "UFO" up in the sky would only be visible to those on the ground not too far from the source laser—but not to those, say, a mile away. The wavelength (color) of the laser's light, the power of the laser, and atmospheric conditions would probably determine just how broad or how narrow such a cone of visibility for observers would be.

The sodium LGS for use in adaptive optics to correct for atmospheric distortions was invented by Princeton physicist Will Happer in 1982 as part of the Strategic Defense Initiative (SDI) but it was classified at the time. Ufologists should be aware of the ability of modern lasers to produce such artificial stars—or, maybe, false UFOs—which can be held stationary or else moved in a smooth path across the night sky. Observers who are unaware can easily mistake such an artificial star for an orbiting satellite or else see it as a UFO—whether that was the intention or not. A whole variety of laser uses and techniques have certainly been of interest to the US military for 40 years or more.

A single bright light in the night sky that is produced in this way is one form of what I term a Laser Projected Illusion (LPI). The image or light in the sky created by a laser may appear to be a luminous physical object moving independently across the sky, i.e. a "UFO" in whatever sense that acronym is understood. It is a potential illusion, whether the person with the laser intends to deceive any observers or not, but nonetheless it is an illusion. A single laser could produce a basic UFO illusion but several parallel lasers connected, or strapped together, could be used to produce a quite impressive LPI that might look like a structured craft with lights attached to it. Such an LPI illusion is quite different from a hologram. That kind of 3-D illusion is something more sophisticated but can also be produced by lasers.

UFOS THAT DON'T BEHAVE LIKE PHYSICAL STRUCTURED CRAFT

Whatever this was it definitely wasn't something from here. It was not any craft from this Earth . . . .

Such declarations by witnesses of what were called "Flying Triangles" or Unidentified Flying Triangles (**UFTs**) during the 1980s and 1990s were not uncommon and did sound sincere. Some said they saw what appeared to be an aerial vehicle moving or stationary in the night sky with lights attached to it. Often they would see three lights which appeared to be positioned at the apexes of a delta wing or a triangle. This kind of UFO was almost always said to be completely silent. Since the lights usually moved together as one, the impression was that it might be some unusual unidentifiable aircraft—or else, as some thought—could the object be an ET alien spacecraft?

Many ufologists and UFO believers who thought the Roswell Incident must have been an alien spacecraft that crashed in New Mexico in 1947 concluded this new kind of UFO would at last provide definite evidence of structured ET craft visiting our planet from beyond. That was the holy grail of ufology but the UFTs proved as elusive as ever.

From 1982 to about 1987 there was a major UFO flap involving

UFTs of this description in the Hudson Valley north of New York City. They were mostly seen at night in upstate New York or in Connecticut. There was also a similar outbreak of UFT sightings by various witnesses in Belgium during the period November 1989— April 1990. On at least one occasion Belgian Air Force jets were said to have been scrambled to pursue such UFOs—though whatever the pilots really did chase may not have been these UFTs. Blips of unknown origin on airborne radar which moved at high speed were thought to be UFTs but that was never confirmed visually by any Belgian Air Force pilot.

Some observers of UFTs in the Hudson Valley flap said they had never seen anything like this before and insisted the objects they saw must be extraterrestrial spacecraft. Some of the witnesses seemed to assume there was an unseen dark body or fuselage from which lights shone out—or else to which the lights were attached. However there was no evidence for the truth of that. An alternative explanation was that some of the moving lights in the sky could have been LPIs projected from the ground below. When such witnesses saw lights in a triangular formation in the sky these may have been produced, in some cases, by lasers attached together in small triangular assemblies projecting the lights up into the sky.

I suggest that some of the UFTs that were seen during the Hudson Valley UFO flap of the 1980s and also some of those seen over Belgium during the 1989-1990 flap could have been laser projected illusions of this kind. Obviously it's not a blanket explanation for all the UFTs that were seen back then but it might account for quite a few of them. That explanation begs the question of who were the perpetrators of such LPI but let's consider that aspect later.

There is no doubt that single lasers capable of producing a small disk of light high in the night sky—whether a clear sky or otherwise— were available before 1980. Observers unaware of what such moving luminosities were could easily think they were actual luminous physical objects—or maybe lights on an unidentified aerial vehicle.

Whether stationary or moving across the sky, such sightings might well be reported as UFOs unless, of course, witnesses said the lights in the sky were something more ordinary like airplanes or satellites.

With any such UFO report the "object" was clearly unidentified and mysterious but, if it was a light from an earthbound laser, it was neither flying in the true sense, nor was it a physical object. Unless an observer knows about a possible laser producing the "UFO" in the night sky he may well believe the intelligence that is directing it is up there too—either a human pilot or possibly an ET alien.

The assumption that the directing intelligence is actually the object itself or else in it, as with that red dot in the forest in Maui being some weird luminous animal, does seem to be a natural one. Only in more recent times do we think that an unidentified object in the sky might be a drone that's being remotely controlled by a person down below.

Let us look at some of the characteristics of UFOs/UFTs that were reported in certain parts of the world during the 1980s and 1990s:

(1) These UFOs were always seen at night or at dusk and were never visible during daylight hours.

(2) The UFOs almost always appeared to move completely silently in the sky above whether seen to be travelling fast or slow.

(3) The UFOs were mostly seen moving slowly across the sky or sometimes "hovering" at an undetermined height. Then, sometimes, they would either vanish in situ or else appear to shoot away, often appearing to zoom directly upwards at "speed impossible".

(4) There was never any sign of how such UFOs were propelled or indeed whether they had any means of propulsion at all. If they really were physical objects, their propulsion systems were totally unlike a jet or rocket engine. Huge acceleration to extreme velocities in just a second or less leads one to believe that, if they

were physical objects at all, they must have been without any inertia. That means that they must have had zero mass.

(5) These UFOs usually appeared to consist of a number of lights, often 3 or more, shining out from—or attached to—a dark fuselage. If there really was a dark fuselage behind the lights it was certainly not visible to any observer and it may not have existed at all. But since the component lights always moved together as one, it appeared as if they were fixed onto or in some unseen rigid frame.

(6) Such UFOs were often referred to as (Unidentified) Flying Triangles (UFTs or FTs) but they also appeared in different shapes such as "delta", V-shaped, "flying wing" and sometimes rectangular.

(7) Sometimes these UFOs would display different colored lights, perhaps red, white, orange, green or blue. It was claimed some of the lights would switch off and other colored lights might then turn on. On some occasions it was reported that all the lights had gone off at once and that made it look as if the UFO had vanished in situ.

(8) Some reports said a thin beam of light would briefly shine down at the ground from such a UFT. Other reports claimed that one, or maybe two, "balls of light" would detach from a UFT, circle around it nearby in the sky or else go down to the ground, and then apparently rejoin the UFT and disappear.

(9) These UFOs were sometimes seen to make flat turns— without banking like an aircraft would. Instant right angle turns without any change of speed were also reported in a few cases. Triangular UFOs sometimes appeared to fly with one flat side—not an apex—leading.

(10) These UFOs were never detected on radar so far as is known. The one or two reports of them being tracked on radar in Belgium and locked onto by fighter planes sent to intercept them may not even relate to the same objects. There were

certainly never reports of UFTs landing or leaving behind landing marks or ground traces. This is one further reason for thinking most UFTs weren't actual physical objects.

None of the ten UFO characteristics listed above indicates that the objects were physical craft moving through the atmosphere. If they had been physical craft they would at least show some aerodynamic properties like aircraft, missiles or any other material object that moves through the air. Nor did these UFOs appear to be acted on by gravity in any way. They gave no indication of having any mass and so a reasonable assumption would be they were not real physical objects. One could in fact say the UFTs had no more substance than mirages, sundogs, auroras, ball lightning (i.e. plasma), or rainbows. Indeed many of them could have been illusions of the kind I call LPIs.

One simple variety of LPI would be the "Flying Triangle" UFO or UFT and the effect could be produced by taping three lasers together in parallel and pointing the assembly at the sky. Best results would be achieved by using more powerful lasers than my little 5mW red laser pointer. Depending on what color laser is used, an apparent "object" in the sky would either have a thin visible beam going up from a laser or, more usually, no beam visible at all. A LPI in the sky might appear to have three lights as the apexes of an isosceles or an equilateral triangle. These would move together like the lights on an aircraft seen at night or the LPI could remain stationary in the sky.

Some of the UFT reports from the 1980s—and especially a few from Belgium—spoke of a thin laser-like beam of light shooting down from the "object" in the sky towards the ground. If any UFT in the sky was in fact a LPI, it then becomes plain that any such visible beam must go up to it from a laser on the ground rather than down from it. It is, of course, impossible for observers to say which way a beam travels when a laser is turned on, since it is instantly visible or not visible.

Many UFOs have been reported with this description and the illusion that this is some kind of physical craft high up in the sky is entirely understandable. Some reports of such UFTs have indicated that they appeared to be huge and estimates such as "the size of a football pitch" were not uncommon. Witnesses were usually unable to say how high up such apparent craft flew but tended to suggest they were somewhere between 100 ft and 2000 ft up. Of course, if these UFOs really were LPIs—rather than physical aerial craft—their estimated size and altitude would be illusory too. It would be rather like an observer trying to estimate the height and span of a rainbow.

A LPI with a basic triangular pattern could be enhanced by attaching additional lasers along just two edges of a triangular frame to produce a "flying wing" configuration. Three lasers attached to the apexes of such a rigid frame could be angled slightly outwards, rather than fixed parallel, so as to produce a larger apparent triangle in the night sky. Or, a LPI projector could be fashioned like a square or a circle with a suitably rigid frame to produce other possible shapes in the sky. The shape of any laser illusion seen in the sky would obviously depend on the configuration of whatever laser cluster was used down below.

An alternative method of producing a laser illusion in the sky would be to use a single laser with a shaped aperture. It would enable the image of a flying triangle to be projected or even some recognizable shape like Batman's famous bat signal! Whatever LPI is projected into the night sky the success of the illusion achieved will depend very much on atmospheric conditions at the time. These and the color of lasers which are used for LPI will determine the visibility of what is seen and also whether or not there is light scattering rendering the laser beam itself visible.

Of course many observers of such an LPI could be totally unaware of the fact there was no real physical craft up in the sky and unaware too of who was responsible for the apparition.

Where military men have experimented with LPIs in attempting to find out how enemy combatants would respond, one assumes they would only try it out on other soldiers rather than civilians. However, some reports of flying triangle UFOs which could have been LPIs were seen in the vicinity of US military bases during the 1980s and 1990s. Equally, it seems just as likely some private individuals who owned lasers—which by the 1990s were becoming available at more reasonable prices—could have produced LPIs simply for a bit of fun. The words 'hoax' and 'prank' spring to mind if that was the case.

I am certainly not suggesting that what I call LPIs are a blanket explanation for all UFOs and/or UFTs reported during the 1980s and 1990s. UFOs which were tracked on radar could not be explained in this way. The intended point of most LPI displays is that the actual source of the fake UFOs remains unseen and unknown so witnesses don't realize what these things are. We can only guess at the intent of people who deliberately project LPIs into the sky above. Some are probably just UFO hoaxers and others may be trying to trigger a wave of UFO sightings for their own particular satisfaction. Who can tell?

Some people insist these phantom UFOs must have a solid fuselage to which the lights are attached or from which they shine out since there are reports claiming the stars in the night sky are blotted out when such a UFO passes overhead. That is not always the case and there have also been some reports of UFTs through which the background stars are visible. Certainly, if the lights at the apexes of such an LPI triangle are comparatively bright they could produce the further illusion of an unseen dark body blotting out fainter stars behind. Some of the UFT sightings during the 1980s UFO wave in Belgium were thought by witnesses to have a thickness of perhaps 8-10 ft but that too could have been part of the illusion.

The ten UFT characteristics listed earlier give a clear indication that many supposed UFOs are not generally speaking physical objects in the sky. Without any mass or inertia that would place them in the realm of virtual or abstract objects like mirages and rainbows. And, very possibly, what I now suggest could well have been LPIs.

If one is willing to accept that a UFT could be a LPI or similar then that avoids the wild speculation of many UFO believers that a supposedly material UFO must have attributes of invisibility, anti-gravity capability, or a fantastic ability to accelerate in hyperspace, perhaps passing into a "wormhole" and vanishing into space-time. But then there are some who believe so strongly that UFOs are of extraterrestrial origin they will go to almost any length to prove UFOs and UFTs are indeed craft with definite physical reality. These people include some who have faked photographs of UFOs showing a dark body or fuselage behind the UFO's lights. Most alleged UFO photos which show this sort of detail are extremely suspect, as we shall see in the next chapter.

COULD SOME "FLYING TRIANGLES" HAVE REALLY BEEN LPIS ?

There are some photographs of these apparent nocturnal UFOs with bright lights around their perimeters such as those shown in the book *Night Siege - the Hudson Valley UFO Sightings* which was first published in 1987 by "Dr J. Allen Hynek and Philip J Imbrogno, with Bob Pratt". Most of the photos in the book just show simply a bunch of lights in the night sky. Only a few drawings by the witnesses of such UFO/UFTs gave any indication of a dark body behind the lights.

That is possibly what a sincere UFO witness might draw if he believed he was really seeing an actual physical aerial craft. However, most of the drawings of the alleged aerial craft are artists' impressions rather than being drawn by actual witnesses. A case in question is a drawing of a huge UFT over Belgium seen among the accompanying pictures.

A particular UFT photograph taken during the Belgian

UFO wave of 1989/1990 did indeed show a black triangular shape behind the three blazing lights at its apexes. It caused several UFO researchers great excitement and was for a time considered best proof of this UFT's physical reality. However, the photo was later conclusively admitted to have been faked and other similar photos supposedly showing any dark body behind UFT lights were most likely faked too.

It is worth noting here that the book *Night Siege - the Hudson Valley UFO Sightings* contained much false information and was an egregious example of what I call **FMWCs** (False Multiple Witness Claims). The promotional blurb for the second edition (1998) ran the line: *Over 5,000 sightings in the last five years...* and it then asked whether all these people could be mistaken? Next, one was told:

> *The "Westchester Boomerang'" was a UFO reported by hundreds of people in New York State and Connecticut between 1983 and 1986 and described by most witnesses as a hovering, immense V-shaped series of flashing lights connected by a dark structure. UFO investigators Imbrogno and Pratt write in detail about the "close encounters" of some 900 people who filled out "witness forms."*

Another red flag as regards this book is the fact that the respected UFO investigator Dr J. Allen Hynek, who gets first billing as one of its three authors, died in April 1986 a year before its first publication. It is unlikely he made any contribution at all to what was published in 1987 and several of the accounts of close encounters with this alleged enormous structured "flying boomerang" were from pseudonymous or anonymous witnesses whose accounts were almost certainly fictional. The writer of the accounts, most likely Imbrogno, has sought in almost every case to emphasize there was a black fuselage, perhaps of gunmetal color or darker, behind the bright multi-colored lights on the UFO that was reportedly seen. When the UFO/UFT lights all went out, this dark aerial craft was said to have become

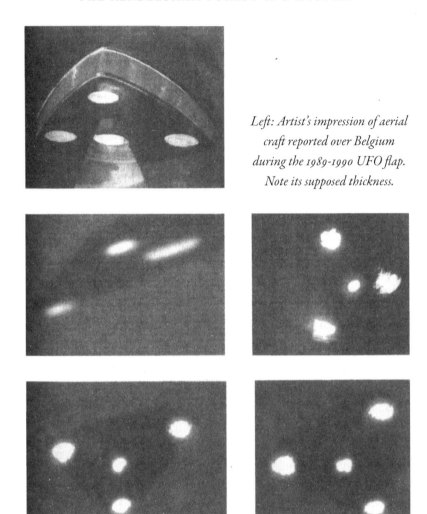

Left: Artist's impression of aerial craft reported over Belgium during the 1989-1990 UFO flap. Note its supposed thickness.

Of the 4 UFO photos above, that top right is definitely known to have been faked. The other 3 are probably fakes too. Most alleged UFT photos merely show lights in the sky with no dark body behind them.

virtually invisible and apparently it then vanished in situ.

If the Hudson Valley UFO flap was greatly exaggerated and promoted by various FMWCs, could some local UFO enthusiasts themselves, like Imbrogno, have been among those using LPIs to encourage it? It does seem possible, but no doubt there were also misidentifications of aircraft lights and the like that could have triggered such a UFO flap.

It does seem very unlikely that the US military would have had anything to do with faking UFOs by using LPIs in the Hudson Valley area at the time, or indeed during the 1989-1990 Belgium UFO wave. Nor does it seem likely that other hoaxers—such as crop circle makers of the 1980s & 1990s—could have been those who were responsible.

If some of the Hudson Valley UFOs/UFTs really were projected laser illusions, the chief suspects would have to be some of the people who fostered UFO groups or UFO cults in that part of the US at that time. During the 1980s interest and belief in ET UFOs had risen to an all time high. The long forgotten Roswell Incident of 1947 had been resurrected and was being claimed to have been the crash of an alien spaceship in the desert in New Mexico. That story became the subject of several books and mushroomed further with claims that small ET aliens—both dead and alive—had been recovered from these crashed UFOs. Several people claiming to have been witnesses at Roswell and at other supposed UFO crashes and/or sightings then came forward. Next there followed, during the 1980s, claims of alien abductions, underground alien bases, cattle mutilations by the aliens in their UFOs, the clandestine retrieval of crashed UFOs by the US military—and even some claims the US government had entered into a secret pact with the aliens!

One group that strongly promoted UFO belief and the fact that Earth was being visited by extraterrestrials was Steven Greer's Center for the Study of Extraterrestrial Intelligence (CSETI). Its stated aim was to initiate contact with the aliens in what was called "CE5 contact—Close Encounters of the Fifth Kind". New members of CSETI signed up for training expeditions in remote places and the practicing of Greer's protocols for attracting and communicating with the aliens which he himself was supposedly in contact with on a telepathic basis.

Apparently joining CSETI was not cheap. But there were many UFO believers who did just that since this was supposedly one guaranteed way to meet ETs. For their training and the necessary protocols for CE5 contact they would go out into the desert and shine lasers or extremely bright searchlights into the sky to draw down the UFOs. Some CSETI people, who had undoubtedly watched the movie *Close Encounters of the Third Kind,* actually made agreements and signed papers to the effect they would, if allowed, travel with the aliens back to the planets they were from and they didn't wish to return to Earth.

A friend of mine from Texas who was not part of the group was once present at Gulf Breeze, FL, when CSETI members met and went out with a view to contacting the ETs. This was in July 1991 and Steven Greer had told them that a UFO would arrive in the sky at 8.02 pm, soon after sunset. The group was well prepared and they waited near the shoreline with huge expectation. The excitement was palpable.

Sure enough, and exactly on time, three star-like points of light appeared in the darkening sky. The three lights, which formed an equilateral triangle, moved as one across the sky towards the waiting CSETI observers. There were cries of amazement and delight and one man could be heard saying "Oh my God, oh my God . . . . This is it!"

Steven Greer, who had brought some large binoculars, peered at the approaching UFO and was telling his people he could see a structured craft. "It's going to land, it's going to land..." The CSETI people went wild with excitement but unfortunately—or maybe fortunately for them!—this silent "UFO" passed on overhead and was eventually lost to sight. Nevertheless Greer had taken video footage of the UFO transit and it was later shown at several different UFO conferences.

During this dramatic CSETI gathering my friend from Texas had also watched the supposed alien UFO through his binoculars. There was no sign at all of it being a "structured craft" and what he saw was simply three points of light moving in unison. There was no visible dark body behind the lights—nor any visible fuselage to which these lights were attached. The lights were quite widely spaced and looked as if they were just flying in formation rather than connected. At what height were they flying? Hard to say. It was puzzling, and it took quite a few years before I found out what this "UFO" must have been.

It didn't appear to be a LPI generated by someone in a concealed position on the ground. Although I never saw this, I was told it looked like three unidentified lights up there moving together in formation. In fact these were artificial satellites about 650 miles up in very similar earth orbits separated by only 30 miles. The Naval Ocean Surveillance System (NOSS) is a series of signals intelligence satellites that have conducted electronic signals intelligence for the U.S. Navy since the early 1970s. They were launched into space on Titan IV missiles from Vandenberg Air Force Base, California, and usually operated in clusters of three in orbit for the purpose of geolocation of Soviet Navy vessels during the Cold War. For years this was a highly classified program whose purpose and existence was unknown to the public. Even so, orbital details and transit times would have been available from various institutions which kept track of most orbiting satellites.

Three NOSS satellites, known as a NOSS satellite trio, passing overhead among the stars is a pretty unusual sight to catch unless one is looking for it and one knows it's coming over. One imagines Steven Greer did know about this NOSS trio transit in advance and where to find such information at a time before satellite transit times became generally available on the internet. It looks a strange sight even to amateur astronomers if one hasn't seen a NOSS trio before. During the 1980s and 1990s there must have been quite a few folks who looked up and, seeing such a triangle, must have thought it a UFT.

Quite apart from NOSS satellites, the subject of LPI fake UFOs could be endlessly debated. LPI does not necessarily apply to what Colonel Halt and the other USAF airmen saw over Rendlesham Forest during the early hours of December 28th 1980 and a simple laser show could easily account for most of the lights that were seen there at the time.

It should be emphasized that an LPI could appear as just a single luminosity moving across the sky. Or, alternatively, it could look like an array of several points of light moving in unison. Although it might appear to be a solid physical craft moving in the night sky, it would really have no more substance than other optical effects like mirages or the luminosity of auroras. LPIs, Laser Projected Illusions, may well have deceived hundreds of unsuspecting witnesses during the 1980s and 1990s into thinking that they were seeing UFOs or UFTs.

PART III

THE BOILERPLATE APOLLO COMMAND MODULES

The Apollo Lunar Mission program of the 1960s culminated when Neil Armstrong and Buzz Aldrin landed on the moon's surface on July 20th 1969. It was seven years earlier President John F Kennedy had declared "We choose to go to the moon" in his famous speech to a large crowd at Rice Stadium in Houston, Texas, made on September 12th 1962. He then pledged that this would be achieved before the end of the decade. Sadly Kennedy never lived to see the success of the Apollo program. He was the victim of a cruel political assassin, Lee Harvey Oswald, in Dallas, Texas on November 22nd 1963.

The ensuing intensive program implemented by NASA concentrated on the development and testing of the Saturn rocket to carry men to the moon and the training of the Apollo astronauts. NASA's Saturn V rocket was the largest and most powerful rocket ever developed. Attached to its top were the three spacecraft components of Apollo that would take the astronauts there, carry them down to the moon's surface, and later return them to Earth and splashdown in the ocean.

Most important of the three components was the Command Module (CM) in which the three-man crew travelled and which provided their life support system. It was attached to the large Service Module (SM) before the start of every lunar mission and the whole assembly was known as the Command and Service Module

(CSM). The third Apollo component was the Lunar (Excursion) Module (LM) which was made to dock with the CSM after Apollo reached earth orbit. The LM was detached from the CSM to take two of the crew down to the moon's surface when the mission had reached the moon and was orbiting it.

When an LM and its crew had completed their excursion to the lunar surface they would blast off in the upper section of the LM leaving its four-legged lower part behind. They would rejoin the CSM in lunar orbit and climb back through the docking tunnel into the CM before heading back to Earth. Before the required rocket burn for that, the LM would detach from the CSM and be crashed on the lunar surface.

When the CM with its three man crew drew closer to Earth, it would jettison the SM before starting its re-entry into the atmosphere.

So, at the end of each Apollo mission, the CM with its three man crew was the only component to be recovered. The SM, without a heat shield, was left to burn up on re-entering the Earth's atmosphere.

Well before the first Apollo moon flights the all important Command Module was tested in every possible way, both as a life support system and as an unmanned payload that was fired into space by rockets. Recovery of the CMs floating in the ocean was practiced both by using helicopters or by winching them up onto ships. The astronauts had to practice egress from Apollo CMs floating in the sea and also had to know how to exit a CM if it had come down on the land by mistake.

Many of the CMs that were used for training and testing were the so-called boilerplate (BP) CMs. A boilerplate spacecraft was also called a mass simulator and it was basically a non-functional craft of the same size, shape and weight as the actual spacecraft. This means it can be use as a simulated payload on rocket launch vehicle flights or for practicing recovery operations at sea.

Above: Apollo boilerplate CM 1102A with functioning crew hatch open.
Below: Apollo Mission insignia.

About 65 Apollo boilerplate CMs were produced for a variety of testing functions and in several differing configurations. They were all the same size and shape which was basically a truncated cone, technically known as a frustum. The frustum was 10 ft 7 in. tall, 12 ft 10 in. diameter (roughly 3.2 metres by 3.9 metres). Weight would be between 4,000 lbs and 9,000 lbs (up to about 4 tons) depending on what the BP was to be used for.

The basic BP Apollo CM was a hollow steel shell fitted with either a fixed steel hatch or a functioning hinged hatch that opened and closed like that on a real Apollo CM. The hinged hatch allowed astronauts to enter the CM before the start of an Apollo mission or to exit from the CM after splashdown in the ocean at mission's end. When astronauts had to transfer from the CM to the LM prior to the LM's descent onto the moon's surface, they would go through a short docking tunnel (of 32-inch diam) at the CM's top end when the modules were connected.

The gross weight of a boilerplate CM would depend on how much of the regular Apollo equipment and instrumentation was fitted. It might include a power source, such as batteries, if needed for a radio beacon and a flashing light used to facilitate ocean recovery.

Several of these 65 BP CMs have ended up in museums or on display at Air Force bases. One is in the Kennedy Space Center visitor complex and another, curiously, sits outside a Dairy Queen in Pennsylvania. Some were scrapped—or supposedly so—and a few are simply missing. A rather incomplete history of Apollo BP CMs can be found on some internet websites such as www.collectspace.com

The particular BP CM that seems to have played a central role in the Rendlesham Forest Incident was designated BP-1206. We will later see what was said about this on the CollectSpace website in December 2010 when the *Daily Telegraph* story first came out suggesting the mysterious Rendlesham UFO was a (boilerplate) Apollo "moonpod" that had been dropped in the forest. However, we need to consider why an Apollo CM was ever at RAF Woodbridge in the first place.

HOW COULD AN APOLLO CM BE THE RENDLESHAM UFO?

What would such an Apollo CM be doing at the twin bases of RAF Bentwaters/Woodbridge in the UK thousands of miles from Cape Canaveral and from the Pacific Ocean? A perfectly valid explanation for that was that there were contingency plans at the time of the lunar missions to rescue and recover Apollo CMs and their crews if they returned from space coming down in some part of the world that was not planned.

One USAF unit that was based at RAF Woodbridge was the 67th ARRS (Aerospace Rescue and Recovery Squadron). Among its functions was to prepare for the possible recovery of returning astronauts who had been on the Apollo or the Skylab missions. The 67th ARRS trained for such rescue missions with their pilots flying helicopters that could pluck an Apollo boilerplate CM from the ocean. The particular BP CM (BP-1206) used for training purposes at RAF Woodbridge was the same size, shape and weight as a fully functional Apollo CM, weighing about 9,000 lbs. It was the only BP CM ever sent to the USAF base at Woodbridge and indeed the only one ever sent to the UK. As things turned out, 67th ARRS never had an opportunity to rescue a manned Apollo CM during its whole tenure at Woodbridge beginning in 1970.

So why should an Apollo CM (boilerplate or otherwise) be seen flown slung beneath a helicopter, presumably by 67th ARRS pilots, near RAF Woodbridge in December 1980? The Apollo moon missions ended in December 1972 and the last Skylab mission ended in 1974. So could what the forester's wife at Folly House saw slung under a helicopter on December 25th 1980 in fact be the boilerplate CM BP-1206?

Nick Pope in his book *Encounter in Rendlesham Forest* (ERF) examines the *Daily Telegraph* story of the roughly triangular "moon pod" seen slung under an American helicopter shortly before the UFO incident that took place in Rendlesham Forest. He says that the 67th ARRS evidently still had their one Apollo boilerplate CM at RAF Woodbridge in 1980 but dismissed both the possibility that it had been dropped accidentally in the forest or even put there as a prank as ludicrous. No arguing with that, but could there be another reason?

Checking further on the internet I found there were two versions of the history of this Apollo BP CM (BP-1206). According to Butch Wilks, who had often visited and worked on the twin US bases during the 1970s and 1980s, BP-1206 was sent back to NASA in the US on a C-141 aircraft in late 1977. He was present at the RAF Woodbridge base when BP-1206 was flown out and he says it was given a great sendoff. There is absolutely no reason to disbelieve that though it appears, as regards the official history of the 67th ARRS many years later, they don't admit it ever left their custody at RAF Woodbridge!

Some of the history of BP-1206 can be found on the CollectSpace website today. As a result of reading the *Daily Telegraph*'s December 20th 2010 "moon pod" story which we referred to earlier, 'Moorouge' (spaceflight historian Eddie Pugh) opened one discussion with:

The curious case of another mishap with an Apollo boilerplate capsule is reported today.

In 1980 there was a UFO sighting in Suffolk that was widely reported to be Britain's most famous.

Now it turns out this may have been BP-1206 which was dropped by a helicopter from the 67th ARRS operating out of RAF Woodbridge. The capsule was dropped in a wood near the airbase as it was being returned from an exercise. It is said that the UFO story was encouraged by the American airmen to cover up the fact there had been an incident.

The capsule was recovered the next day.
What is curious about this is the date. 1980 is well after the ending
of the Apollo flights. Do we have another mystery?

Butch Wilks was quick to reply to Moorouge and obviously keen to dispel any suggestion that the 67th ARRS could have been responsible for the Rendlesham UFO incident. He wrote:

Sorry, not possible. Some background: I visited and worked on the bases for a number of years from 1970 to now and back in late 1977 BP-1206 was sent back to NASA on a C-141. I was on the base at the time and it had a big sendoff.

As a side note, the 67th ARRS got in the first of the shuttle mockups in the first two months of 1978.

Moving on to the 1980's sighting: one more time, I was on the base at the time and it is not possible that a helicopter from the 67th ARRS was flying that night, as we had a ground fog for the days of the sighting and all flights were grounded at the time.

He goes on to say that there were other UFO sightings at RAF Woodbridge/Bentwaters, one as early as 1960, but it was unclear whether Wilks even realized that RFI involved a large conical object down on the ground in the forest which was about the size and shape of BP-1206. He may also have been unaware at the time that this particular BP CM had somehow been brought back to Woodbridge before the end of 1980. In any case, he dismissed the suggestion that BP-1206 could have been dropped in the forest by 67th ARRS. Also, it seems he was the only person to maintain there had been a ground fog which prevented all flying at the time in question. Certainly John Burroughs and Jim Penniston made no mention of fog on the night they encountered the landed UFO in the forest clearing.

Over a month later Moorouge again posted on CollectSpace saying that Apollo boilerplate BP-1206 was returned to the US in 1977 and that as of March 2008 it was located at Patrick AFB, Florida. Its 1977 return to the US was clearly inconsistent with the December 2010 story in the *Daily Telegraph* unless BP-1206 had been secretly flown back again to the UK sometime before 25th December 1980. If so, it must have been done for reasons which were never publicly disclosed.

One thing which did however become clear at the end of 1980 was that the boilerplate CM was back at the base where it could be seen to be parked in early 1981. According to Moorouge, soon after the Rendlesham UFO incident BP-1206 was moved away from its site in front of the 67th hangar to a secluded spot out of the view of the public. Whether officers at the base did that to quell speculation that the boilerplate was involved in the UFO episode or not is obscure. No doubt the twin bases were abuzz with rumors but few of the public in early 1981 knew quite what was said to have gone on. It was only in October 1983 that the British public at large heard about the alleged Rendlesham UFO when the *News of the World* published sensational headlines plus their story implying the UFO was an ET spaceship.

Although the forester and his wife at Folly House had seen what looked like an Apollo CM slung below a helicopter shortly before the Rendlesham Forest event few people knew of this in the weeks that followed and, apart from these two, hardly anyone saw a connection. So, it was a full thirty years later that the *Daily Telegraph* published its story and people started to ask 67th ARRS officers awkward questions in 2010 and 2011. Apparently few 67th ARRS men were willing to talk about it and those that did suggested that—if the UFO had been an Apollo CM dropped by one of their helicopter pilots in the forest— alcohol and/or drugs were possibly involved. It did seem a plausible explanation since it had occurred during the Christmas holiday. So was that the reason for the USAF's 30-year coverup of the story? I suggest it was definitely not the true reason.

The dialog about BP-1206 on CollectSpace continued about ten days after Moorouge's posting that said the BP CM had been returned to the US in 1977. A member of CollectSpace named APG85, quoting what Moorouge had said, riposted:-

The Woodbridge boilerplate was not returned in 1977. I was stationed at Woodbridge from 1985-1992 in the 67ᵗʰ ARRS and "we" had possession of it in that period.

Immediately Moorouge acknowledged that he must have made a mistake. He went on to state that possession of Apollo boilerplate 1206 was transferred to the Smithsonian in title only on 29 April 1976. A year after that a loan agreement between the Smithsonian and 67ᵗʰ ARRS was executed which allowed them to keep custody of it for as long as they still needed it. He says it allowed 67ᵗʰ ARRS "to continue using it as a training vehicle" though, I suggest, that doesn't necessarily mean that was what it was actually used for. The BP CM was never sent back to the Smithsonian in Washington DC at any time, and from July 1977 the 67ᵗʰ ARRS had custody of it both at RAF Woodbridge and in Florida. In June 1991, at the discretion of the USAF and, as subsequently reported to the Smithsonian, BP-1206 was shipped back from RAF Woodbridge to the 71ˢᵗ ARS (Air Rescue Squadron) at Patrick AFB, FL.

I contacted Eddie Pugh to ask whether he was completely sure that BP-1206 had not been flown back to the US in late 1977. He said that he had to accept this as APG85 had insisted the boilerplate never left RAF Woodbridge. Certainly that was the official line of the 67ᵗʰ ARRS after the December 2010 report in the *Daily Telegraph* and, although CollectSpace member APG-85 was not stationed at Woodbridge any time before 1985 this seemed to be what he was lead to understand.

Did Eddie think that Butch Wilks's account of BP-1206's departure from Woodbridge in late 1977 sounded to be a fiction? It sounded perfectly true to me, I said, and it made perfect sense since there was no longer any need for it then as a training vehicle five years after the last Apollo mission in 1972, the last Skylab mission in 1974, and the one and only Apollo-Soyuz mission in July 1975. Eddie said that if BP-1206 had been flown back to the US in 1977 it obviously must have been returned to Woodbridge sometime before the end of 1980. So why not ask Butch Wilks about what he said on CollectSpace?

That indeed is what I have tried to do but I was unable to get any answer. If the 67th ARRS now maintains the boilerplate never left Woodbridge in 1977 and remained there all the time, it's very unlikely he'd contradict their version of the history of BP-1206. Whatever the truth, any involvement of 67th ARRS in a December 1980 Special Ops exercise in Rendlesham Forest MUST be denied—even 40 years after the event. But now that much of the background has become public knowledge, can anyone really be sure of what official sources both inside and outside the US military claim is the truth?

When BP-1206 was returned to NASA in Florida in late 1977 there was no intention of using it further for rescue and recovery training exercises. As far as NASA was concerned then all future US manned spaceflights would use the Space Shuttle which would either land on a long runway at the Kennedy Space Center (KSC) in Florida or at Edwards AFB in California. There were plans too for other possible emergency landing sites in the US and abroad but sea recovery was never even considered possible. So it seems BP-1206 would simply be kept at some USAF base as a rather large memento of the Apollo era. As such, NASA probably had it repainted with an Apollo Mission Symbol and with additional US markings to historically identify it.

In 2011 with members and ex-members of 67th ARRS disputing that BP-1206 was sent back to the US in 1977, it was undeniable that it had been there, in the keeping of 67th ARRS, at Woodbridge during almost all of the 1980s. However there is a puzzling disagreement here and, if it did go back to America in 1977, as seems most likely, one really has to ask the question why and when was it returned to

Woodbridge. I suggest it was secretly sent back to the 67th ARRS in late 1980 with a very special mission in mind.

Sometime after NASA's Apollo and Skylab missions ended during the 1970s, the 67th ARRS was given a totally new direction that was not connected in any way with air-sea rescue and recovery of spacecraft launched by NASA. Its primary mission was to be a covert one that would "provide worldwide clandestine aerial refueling of special operations helicopters" and it was to have a secondary role that included "infiltration, exfiltration and resupply of special operations forces by airdrop or airland tactics". Such operations would be totally secret and could be ordered in many different parts of the world such as the Middle East. The 67th would use its helicopters to take US Army Special Forces—the "Green Berets"—into and out of enemy territory and/or war zones with no questions asked. Maybe they would even carry US Navy SEALs on their black ops missions behind enemy lines in Iraq or Afghanistan. These operations would be very different to practicing for the possible recovery of Apollo CMs. Indeed, such operations gave no plausible reason for sending that Apollo boilerplate CM the whole way back to RAF Woodbridge!

The new covert mission directive given to the 67th ARRS appears to have been inaugurated in 1980 but it certainly wasn't announced or made public knowledge at the time. In fact such knowledge of the squadron's new mission was not made public until several years later and few would make any connection with the Rendlesham Forest UFO incident. To set the 67th ARRS on its new course there would be a new commander to take over at RAF Woodbridge in July 1980.

COL CHARLES WICKER APPOINTED 67TH ARRS COMMANDER

On July 25th 1980 Colonel Charles E Wicker took over as Commander of the 67th ARRS at RAF Woodbridge. His appointment no doubt meant that he was well qualified and well

Above: BP-1206 at RAF Woodbridge

connected with regard to the important new role planned for the squadron. Although the pilots of the 67th were not themselves considered Special Forces they would effectively be on a par with them when involved in covert operations.

The above photo, which was released and became available on the internet, is now displayed at the BCWM at Bentwaters. It is really quite a puzzle. Since Colonel Wicker's name is on the black banner one would assume it must have been taken sometime after July 1980? The photo was evidently intended to make one think the mission of the 67[th]ARRS was, in 1980 and later, the same as ever—Rescue and Recovery of US Apollo spacecraft—that is "the Best in Rescue". We now know that by 1980 the main mission of the 67th was no longer to be rescue and recovery, so was the photo meant to mislead the public by saying the mission was unchanged? It certainly looks as if the 67th ARRS was trying to do just that. Apart from covering up their new cloak and dagger role, the photo was very likely intended to dissociate the squadron from any suspected involvement in the notorious RFI.

The question also arises as to when this photo was taken. We do know that BP-1206 was seen back at RAF Woodbridge very soon after the Rendlesham Forest UFO incident and it was seen and photographed on the base there during 1981 and in following years. But was it ever seen and photographed there between late 1977, when it left, and December 26th 1980 when RFI occurred? I'm sure past and present members of 67th ARRS will assure one it was, but there's no convincing proof of it.

It is quite possible this photo of BP-1206 was taken sometime in the 1970s before the CM's departure for America in 1977. If that's true the black "banner" that is supposedly draped across it must have been added to the old photo years later, in view of Col. C. E. Wicker's name being on it. Looking carefully at the photo one can see the "banner" doesn't appear quite right. It's not painted on the sloping surface of the CM and it doesn't even appear to be taped on since its straight edges don't align with the slope. In my view the original photo seems to have been doctored by addition of the square flat "banner" at some stage before re-publication. Although the text of the banner curves, as if it followed the rounded surface, the photo still looks to have been doctored and unlike how it must have appeared originally. If the photo was in fact taken some time after the RFI, such a banner could also have been used to cover up remaining traces of NASA's Apollo Mission Symbol—if that was still distinguishable on the CM surface.

Whether or not BP-1206 returned for a time to the US in 1977 it was certainly visible sitting in one or two places on the base from early 1981 on. At some stage in early 1981 it reappeared sitting beside the 67th ARRS hangar or the Pararescue (PJ) building. There's an online suggestion that it was moved away from public gaze for a time following the Rendlesham Forest UFO incident in December 1980. Perhaps this was merely an assumption that it had been moved away from public gaze whereas it simply hadn't been visible outside before RFI. When it did reappear a few weeks later people might naturally assume it had been moved and that it must have been sitting there on display before RFI. Of course, former members of 67th ARRS deny it ever left Woodbridge and they usually strongly deny the squadron had anything to do with the UFO incident.

A further fascinating detail is that Col. Charles I Halt knew Col. 'Charley' E Wicker well enough. They were racket ball partners and played the game together on the USAF base. On the day following the UFO incident Halt talked with Wicker about what had supposedly landed in the forest the previous night. Wicker apparently replied "We [the 67th ARRS] were down for the [Christmas] holiday but, if it happens again, call me and I will get a crew airborne". As one of the very few people at the twin bases who almost certainly knew what the RFI really was, there should be little surprise that Charley Wicker affected total ignorance. A military secret like this should NEVER be disclosed—not even to one's wife or one's closest friend!

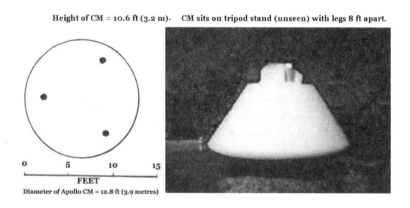

Height of CM = 10.6 ft (3.2 m). CM sits on tripod stand (unseen) with legs 8 ft apart.

FEET

Diameter of Apollo CM = 12.8 ft (3.9 metres)

Above: Diagram showing dimensions of boilerplate Apollo CM.

WHAT DO MEN'S SKETCHES OF RENDLESHAM UFO SHOW?

Unsurprisingly the rough sketches of the craft made by Burroughs and Penniston look somewhat different though both are more or less triangular in shape. Their sketches were made separately and done from memory between one and three days after the encounter in the forest—and apparently not *during* the encounter as some have previously reported. In fact no sketches would have been necessary had any of the photographs of the craft taken by Penniston at the time come out. Penniston was told that all his photos were all overexposed and therefore useless so had been thrown away. Whether there was any truth in that or not, the photographic prints were undoubtedly withheld from Penniston and confiscated by the AFOSI investigators.

Since this encounter with the landed UFO occurred on a December night at 3 am inside a dark forest there was very little ambient light to illuminate the craft. Such light as there was apparently emanated mostly from the craft itself and it seems there were a number of lights of different color on it which turned on or off in a way that did not follow a recognizable pattern. Although the object's general outline, which was roughly triangular was visible it would have been difficult for anyone to draw it as a 3-D object when viewing it from any particular position.

Let's look first at the sketch of the UFO drawn by John Burroughs.

It shows a triangle or cone-shaped object on top of which is fixed an oval shape. This appears to be a light or lights of some kind. He recalls seeing a "red oval, sun-like object in the clearing" but he does not recall seeing any craft. In fact Burroughs, unlike Penniston, has no recollection today of anything that he perceived as a "craft". He certainly has no recollection of seeing the object—this UFO— silently take off, maneuver through the pine trees and shoot away into the sky at "speed impossible" as Penniston claims he did.

Under the triangular object in Burrough's sketch above is drawn

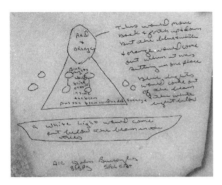

What Burroughs wrote beside the sketch, with a line to its top, was:-

" *This would move back & forth up & down but the blue & white & orange would come out when it was sitting in one place*

blue lights would come out of the beam & the white light below"

[written across the body of the craft is] "blue lights would blink on and off in line   the beam   plus the beam would be red & orange"

[In a shallow dish-like area below the craft he wrote]   a white light would come out below the beam in the trees"

[At the bottom he wrote] "A1C John Burroughs
81 SPS S Pol CF T"

Above: John Burroughs' rough sketch of the craft. The triangular outline of the UFO is surmounted by a large oval object which is presumably the bright white - as well as red/orange - light that was on the top of the craft. This light (or quite possibly two separate lights) was changing, coming on or off in some irregular pattern and moving up and down and sideways too.

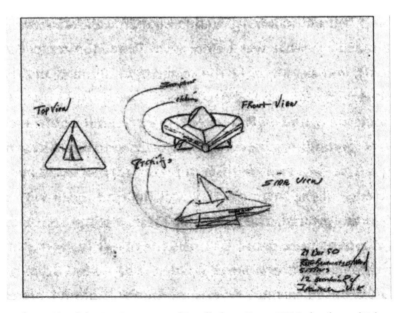

Above: Sketch by Jim Penniston of Rendlesham Forest UFO that he and John Burroughs encountered in the the early hours of December 26th 1980. This was drawn from memory in his police notebook on December 27th at his lodgings in 12 Blenheim Rd, Ipswich, UK.

what seems to be a pool of white light on the forest floor. He said small fixed blue lights in a horizontal line shone from the sides. The white light plus a red/orange one appears to be situated at the UFO's apex.

The sketch by Jim Penniston of the UFO is more detailed than what Burroughs drew but does not show the different lights. It does not mean this supposed configuration of the object is correct and what Burroughs drew was wrong, but it shows Penniston, when he got to draw it, clearly thought of the UFO as some sort of flying craft. He does claim that he touched it and walked around it at least twice but his drawing of the UFO from memory a day or two later makes it appear roughly triangular but not quite like a cone sat on its base. He probably had every reason to think the object was some sort of craft—though possibly not an alien spaceship!

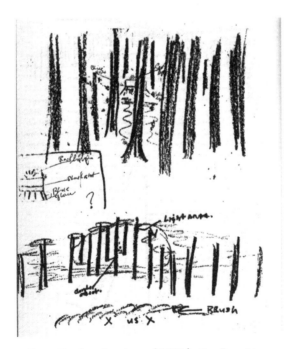

Above: Further drawings of UFO by Jim Penniston

Maybe on reflection he had thought it looked more like some kind of unknown military aircraft?

It has been said that neither Penniston or Burroughs believed at the time there was anything to UFOs or to visitation by extraterrestrials. Apparently that remains very much the case today and we are told that Penniston's thoughts now on his Rendlesham Forest encounter in 1980 are that the UFO may have been time-travellers arriving from sometime in the future. In any case, one might assume back then he thought the craft was an aerial vehicle, possibly a secret military one.

That very likely influenced his sketch of it. One of these shows it with a squat triangle-shaped body with a definite angular nose at what must be the front end. It also has a pointy cockpit-like protrusion on the topside which presumably corresponds to the oval-shaped object or light(s) on top depicted in John Burroughs' drawing. If Penniston's sketch was intended to look like some sort of VTOL aircraft, the only military airplane that it vaguely resembles from 1980 would have been the British Harrier Jump Jet or its US counterpart the McDonnell Douglas AV-8. That VTOL aircraft could land in or take off from a forest clearing and film of it published in about 1980 showed it doing just that.

It's worth noting here that Col Halt's evaluation of Jim Penniston made in his notes some time later was as follows:

Sgt Penniston has a lot to contribute. . . . I think he's holding out to "sell" a story. He is, however, a very competent individual and can be trusted. I'm convinced his story is as he says. He was so shook he had to have a week to recover.

Maybe Penniston's sketches shown above were somewhat influenced by his plan to sell a story? The strange craft which he depicts in these sketches certainly looks rather more like something that could fly under its own power than an Apollo CM. If these sketches by Jim Penniston of the UFO don't look exactly like the shape of a CM, those he made of it in a different drawing with the craft partly hidden behind trees (below) definitely do. In the drawing of his below, a truncated cone (pointing upwards) is topped by what appears to be the squat cylindrical section on which a red light is fixed. There's no indication of when the sketch was made—unlike the previous ones supposedly made on December 27th—but it could have been done when he and Burroughs first approached the UFO from a distance of about 50 metres (164 ft.) In any case, the sketch was said to have been handed over together with the previous ones to his AFOSI/CIA interrogators.

The truncated cone of an Apollo CM would have a roughly triangular outline to an observer. Its dimensions—10 ft 7" (3.23 metres) high by 12 ft 10" (3.91 metres) diameter—are shown in the CM dimensions diagram at end of last chapter. That certainly corresponds quite closely with the estimated 3 metres by 3 metres size of the Rendlesham UFO that Jim Penniston first suggested in 2003.

But one should also take into account the fact that the craft was quite evidently sitting on legs positioned in a tripod arrangement which raised it several inches off the ground. When seen in the dark its rounded cone-like shape might easily be described as triangular. His sketches of it in a notebook were supposedly made a day or two later from memory. One sketch of the craft shows two of the tripod legs sticking out from under it. Halt's report in his memo says the legs–or more likely the foot-pads terminating them—were each 7 in. in diameter and each made a depression of 1 ½ in. depth in the ground.

If the Rendlesham Forest UFO was indeed a boilerplate Apollo CM adapted as a THW, we should consider what the various colored lights might have been that were fitted to it. First there seems to have been a flashing red/orange light fixed to the top of it that was initially turned on. That sounds very much like a regular SAR beacon of the sort that would be used for Search and Rescue exercises at sea.

Secondly there is the description of "blue lights that would blink on and off" in a (horizontal) line across that part of the cone facing towards Burroughs and Penniston. This curious configuration made no sense to me at first but there's a further clue here that persuasively links this to BP-1206. That particular boilerplate, which for years had been in the custody of the 67th ARRS, had a horizontal ring of open holes all the way around it—unlike most other boilerplate CMs. The original purpose of these one-inch holes was to admit seawater to an internal horizontal toroidal chamber in the CM so that the additional weight would give it increased stability in the water after splashdown. Ideally a CM was meant to float in the sea in a 'Stable 1' configuration, meaning with the cone upright, and the extra weight of ingested seawater in the toroidal chamber was meant to keep it that way.

This almost unique ring of holes could have been plugged with small blue lights when the BP CM was repurposed and it would never again have to be dropped in the ocean. Apart from that, a circle of flashing bright blue lights like this would be clearly visible to the crew of a helicopter above when the time came to remove the CM from the forest clearing. It may be no coincidence that the Cash-Landrum UFO, which was described previously in the story of that encounter on December 29th 1980 in Texas, also had an equatorial ring of small blue lights. If both these supposed UFOs were in reality THWs then the rings of blue lights were most likely meant to make them clearly visible to helicopter pilots flying above them on night-time missions.

Lastly, there is the description of a bright white light that shone out intermittently from the apex of the UFO as a broad beam. Burroughs says this beam of this light moved "back and forth, up and down" illuminating the trees above and also the forest floor underneath. Whether the light itself moved up and down in the cylindrical 32-inch docking tunnel (which is positioned vertically in the center of the CM) we can only speculate. If it "came out"

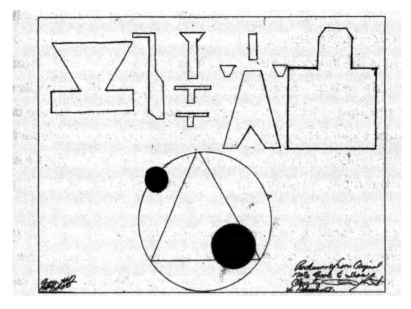

Above: Sketch by Jim Penniston (from his notebook) of the main symbol on the UFO's surface. This bears a striking resemblence to the Apollo Mission symbol. Also the "glyphs" (top) which he said felt rough like sandpaper.

(according to Burroughs) from such a fixed central tube that would render the light itself visible—rather than just its beam which evidently shone both upwards and downwards. If the light's beam moved "back and forth" as we were told, one might also suppose it was able to swivel horizontally and be made to point at anyone approaching it from a particular direction.

This movable white light on the UFO could well have been a Directed Energy Weapon (DEW) capable of delivering a large burst of electro-magnetic energy at any person approaching when required. We shall return later to consideration that the white light may have beamed out from a DEW and it was tried out on the two unsuspecting men of the 81TFW Security Police without their knowledge or their consent.

SYMBOLS REPORTED BY PENNISTON ON CRAFT'S SURFACE

Foremost is that large circle enclosing a triangle with a dark circle at upper left and a larger dark circle at bottom right. That did seem vaguely familiar though admittedly it took quite a time before anyone could put their finger on it. It eventually dawned on some people that this was something very similar to the Apollo Mission symbol.

The match between Penniston's sketched symbol and the Apollo Mission symbol is very striking although they are not quite identical. He had drawn the symbol from memory later rather than when he was on the spot standing next to the spacecraft—though I believe he has subsequently claimed he sketched it at the time he was there. His report stated he took several photographs of the craft when he was close to it. So he would hardly have needed to sketch the symbols then since he could photograph them? However, when eventually he returned to the base and turned his camera in for the photos to be developed, he never got to see the results. As I said earlier, these were almost certainly confiscated by AFOSI investigators and he was told that all his photos had been overexposed and were worthless.

Opposite is a picture that appeared on the internet pointing out the similarity between the Apollo Mission symbol and Penniston's sketch.

The mission symbol shows, on the left, the yellow disc of the Moon bearing an image of the Greek god Apollo on its face. On the right is the blue planet Earth with the continent of North America visible at center. Between Earth and moon stands a large stylized letter A for Apollo. Also, between earth and moon, are seen the stars of the constellation Orion. These are shown with Orion's belt stars along the crossbar of the letter A. Curved white lines between Earth and moon symbolize the translunar trajectory along which three Apollo astronauts on each lunar mission would fly to and from the moon. Around this in a broad ring round the center are the words: APOLLO and NASA.

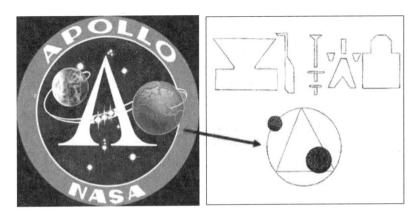

Above: Picture showing similarity between the Apollo Mission symbol and Penniston's sketch.

Here it should be said that the Apollo Mission symbol was not designed for any particular manned lunar mission. All the actual Apollo missions, both the earth orbital and translunar ones had their own individual insignia like a special circular badge but that would not usually have been painted on the outside of a CM. However the general mission symbol as well as the American flag and the words UNITED STATES were sometimes painted on the exterior of the Boilerplate CMs which were in the possession of NASA.

Identification of such boilerplate CMs that floated in the ocean was thought necessary for at least one boilerplate CM (BP-1227) went missing at sea. This happened in the Bay of Biscay in 1969. It was found by a Russian ship and taken back to the Soviet Union. A year later it was returned with great ceremony to the crew of the American Coast Guard icebreaker Southwind when it made an unprecedented courtesy visit to Murmansk despite this being during the Cold War.

But what could one possibly make of this apparent Apollo CM which had, it seems, landed deep in Rendlesham Forest in December 1980 and then just as quickly departed. The supposed Apollo Mission symbol on it gave rise to some extraordinary speculation. Was there a secret Apollo program to the moon—and maybe beyond—being conducted by the US government? Had this CM fallen to earth there after some lengthy secret mission in outer space during which the famous Apollo insignia had faded almost completely away? Was the US government secretly in collusion with ET aliens who were flying in their NASA spacecraft?? Such fevered speculation flared briefly on the internet!

It has been said before that if it looks like a duck, walks like a duck and quacks like a duck, then it most probably is a duck—even if it's found in the most unlikely place. Whatever the unlikelihood, there are many people who now feel quite sure the Rendlesham Forest UFO of December 1980 <u>must</u> have been an Apollo Command Module.

THE "GLYPHS": ERASING THE CM'S IDENTIFYING MARKS & NUMBERS

On July 1st 2008 a boilerplate Apollo CM was placed on a flatbed truck and taken from Patrick AFB Florida along A1A through Cocoa Beach and Cape Canaveral to Excell Coatings at Port Canaveral. It was taken there for refurbishment and external repainting of its insignia and identifying marks. This particular boilerplate, #1206, was one that had been used to train DoD forces in recovery operation procedures for the Apollo and Skylab programs during the 1960s and 1970s. It was, in fact, BP-1206 which had been based with the 67th Aerospace Rescue and Recovery Squadron at RAF Woodbridge, UK.

It was indeed the very boilerplate CM that had been flown back to the US in a C-141 airplane in 1977 when it was no longer needed for recovery training purposes connected with the Apollo program. If it was the BP CM that was used for a secret THW test in Rendlesham Forest in December 1980, its conversion for the purpose at the time would obviously have required removal of all its external identifying markings and this must have been done before it was secretly flown back to RAF Woodbridge. That particular task would have been very different from the repainting and renewal job done by Excell Coatings in 2008. Let us consider the following possible scenario.

Sometime in late 1980 a senior maintenance aircraftman, "Jack", was probably summoned to the Wing Commander's office at an Air Force Base in Florida. Jack would most likely have been told by the Wing Commander "I've got an important job that I want you to do now."

Apparently word had just been received from Washington that they wanted the boilerplate Apollo CM that had been kept at the base since it was returned from England in 1977. Jack knew boilerplates like this one had been sent by NASA to various Air Force units in the US and to US bases abroad for the training of helicopter crews who might be involved in the rescue and recovery of returning astronauts.

Most boilerplate CMs had never been intended to go into space. This one had no working hatch for access to its interior but it carried a flashing red SAR beacon on top powered by heavy duty batteries. The SAR lights were to assist helicopter crews sent out to recover the CM by day or by night and to locate it from afar. It also had a radio beacon to assist with location and recovery at sea.

On the boilerplate CM's exterior was emblazoned the circular Apollo Mission insignia and inside that the words APOLLO and NASA. In addition it carried the words UNITED STATES in bold letters together with an American flag. All of these distinguishing marks had been painted on its smooth metal surface with tough epoxy paint just like identifying emblems and letters on any USAF aircraft.

Although the boilerplate CM had been kept in the corner of a hangar now for about three years Jack was proud of it and he had come to regard it as his own. So, Washington now wanted it back? He was dismayed to hear that a transporter aircraft, probably a C-141, would be arriving at the base to collect it at 0600 the next day.

He asked the Wing Commander what they were going to do with it and why they needed it so quickly. Surely no one was going to send it to the moon! Perhaps it was going to an aerospace museum but that hardly explained Washington's sudden request to come and collect it.

The Wing Commander said that Washington hadn't told him why they wanted it but they had made a further unusual request and that was why he'd summoned Jack. They wanted all identifying marks and every single letter and number painted on the boilerplate completely removed. He said he needed Jack, and Jack alone, to do this task and it had to be done today however long it took.

Jack was aghast. He had intended to finish up early and go home to watch the NFL game on TV. Now he would have to spend hours at a task which was something he'd never done before. "How am I meant to do that?" he asked. "We'll that's something you'll have to figure out", said the commander. "Don't ask anyone else to help you and don't tell anyone what you are doing in the hangar. "The order from Washington is classified, so get on with it and tell nobody."

The only way Jack could think of to remove the insignia from the BP CM was with a Black and Decker power sander and/or an electric drill fitted with an oscillating wire brush. Whichever method he used would produce a horrible result—unless perhaps he masked off all the areas outside the markings he was trying to erase. So, using a rotatory wire brush, he found he was able to obliterate APOLLO and NASA and everything else inside the big circle except the very persistent 'Moon' disk and the 'Earth' disk—plus the triangle that had once been a big letter **A** for Apollo. Other markings such as the US flag he enclosed with straight lengths of duck tape and then, using a power sander, he removed all the lettering plus serial numbers and suchlike.

Above: Identifying markings such as this one were painted on some of the BP CMs. Penniston noticed a roughened "glyph" of this shape on the surface of the Rendlesham UFO. 3M duck tape as shown was probably used to mark edges of the area to be erased before sanding it.

Jack never did get to watch the football on TV and he spent over half the night trying to complete his task. When the transporter aircraft flew in at an early hour to collect the BP CM little was left of its Apollo Mission symbol or any letters that could identify it as a CM. Various roughened oddly-shaped areas were visible where the identifiers or letters had once been. These were the areas subsequently described by Jim Penniston as "glyphs". And, years after 1980, the description of the remaining outline of Earth and Moon—which Jack had failed to obliterate—would allow identification of the strange object which had mysteriously appeared in Rendlesham Forest in December 1980.

If my suggestion above that the hypothetical "Jack" spent hours scouring out insignia and markings on a boilerplate Apollo CM before it was taken away to be further modified—and soon after that flown to England—is close to the truth, then the identity of the

Rendlesham UFO becomes clear. It was a boilerplate Apollo CM repurposed as a THW by a US intelligence agency for use by US Special Forces.

Penniston says the surface of the craft was "hard and smooth". It was "warm to the touch and felt like metal." Referring to the "glyphs", as he called them, "they were nothing like the rest of the craft, they were rough, like running my fingers over sandpaper". That indeed is exactly how they would have felt if our hypothetical Jack had sanded away all Apollo identification markings in just those areas.

Penniston sketched the "glyphs" all grouped together rather than in any positions relative to each other and to the Apollo Mission symbol. The suggestion that they were like alien hieroglyphs with some definite meaning just has to be wrong. It wasn't until years later when the truth dawned that the "glyphs" were areas where markings had been intentionally removed, or obscured, did Penniston's description of the exterior of the Rendlesham UFO make any sense.

Unless this "UFO" in the forest really was a re-purposed BP Apollo CM any interpretation of the roughened "glyphs" is very difficult indeed. When one realizes the "glyphs" must have resulted from almost complete removal of identification marks by our hypothetical Jack—the whole thing then begins to fit and I seriously suggest that an unbiased observer will accept this is the solution to the enigma.

It's doubtful that Jack ever knew what became of the boilerplate CM which was taken away from him. Or whether he ever saw Penniston's sketches which were eventually revealed to the public some 23 years after the encounter in Rendlesham Forest. Any such THW with a similar purpose today would undoubtedly be far more sophisticated and certainly more hi-tech. It would likely be some kind of drone but today's hostage takers and terrorists are probably far more savvy and it's unlikely they would be fooled by a drone disguised as a UFO.

A HIGHLY DUBIOUS STORY OF UFO'S LIFTOFF AND DEPARTURE

Everyone who has examined accounts of RFI by the USAF witnesses—and the primary witnesses in particular—will have wondered just how much of each account is true. Is this witness account 100% true or is it, maybe, 90% true and 10% false? Some obviously consider some of the accounts 100% false but I feel sure this is not the case. To make a fair judgement on each particular account the best thing that one can do is to identify parts of it which appear quite definitely false and reject those before examining the rest of it.

That parts of some accounts are false can hardly be disputed since, when examined together, accounts are often contradictory. The contradictions may be due to omission of pieces of the overall story or they may be due to embellishment or extrapolation of what actually happened. With the "sanitized" versions of the witness accounts that were released by the USAF authorities soon after RFI there is no mention of Burroughs and Penniston's encounter with a landed physical craft in the forest. There's obviously mention of "unexplained lights" in the forest and how these were investigated by the US airmen who walked a long way into the forest to look. Later accounts by the same witnesses reveal a lot more took place there.

Now here is something which is patently untrue:

> "In my logbook (that I have right here) I wrote: "Speed—impossible."
> Over eighty Air Force personnel, all trained observers assigned to the Eighty-First Security Police Squadron, witnessed the takeoff."

This is from Jim Penniston's position statement (quoted earlier in this book) which he produced for the November 12th 2007 press conference on the UFO subject at the National Press Club in Washington, DC. It is repeated in the 2014 book *Encounter in Rendlesham Forest*, co-authored by Penniston, and so presumably he stands by this claim. I did query if the first instance of the word "eighty" above might have been a typo for "eight". I was told by Nick Pope that it was not and that "eighty" is what Penniston maintained.

Over eighty AF personnel all trained observers? Really?? All of the other accounts point to no more than five men out in the forest at the time including Burroughs, Penniston and Cabansag. Where were the rest of them who allegedly witnessed the takeoff and what have they got to say today? Since the landed UFO was a mile from East Gate, deep in the forest, how could anyone at RAF Woodbridge, or for that matter at RAF Bentwaters three miles away, have possibly seen the alleged takeoff? Quite apart from that, it's highly improbable 80 men from the 81st Security Police Squadron were even on duty then (0300 on December 26th). Surely this claim of 80 witnesses is preposterous!

Penniston's 2007 position statement seems to have been prepared very much with the UFO community folk at the press conference in mind. Physical alien spacecraft were often said to silently hover and then shoot off at unimaginable speed. There are various other claims of UFO encounters where this is said to have happened and there were said to be "dozens of witnesses". It was sometimes claimed there were multiple witnesses to the alleged flying saucer crash at Roswell in 1947 but later investigators were never able to find *any* reliable witnesses to that event when they started looking for them.

This sort of "False Multiple Witness Claim" (FMWC, as I call it) is sometimes found in UFO literature and many UFO researchers have been taken in by such claims when they failed to investigate further. Besides many alleged claims of non-existent multiple witnesses at Roswell, there was a similar FMWC by an alleged UFO contactee who said an alien UFO had landed in a public park near Denver and it was witnessed by dozens of people gathered there on a sunny afternoon. Needless to say the claim was untrue and gullible investigators could easily have found that out if only they'd tried to locate any witnesses.

If the alleged multiple witnesses to the Rendlesham UFO takeoff are a false claim, one should ask too whether the alleged takeoff itself actually happened as Jim Penniston described it. Or was that too a sensational embellishment to an otherwise true story? If we accept that there was an actual physical craft sitting there in a clearing in Rendlesham Forest for quite some time between 0000 and 0300 on December 26th, whatever became of it and how did it actually leave?

Burroughs saw the lights in the forest and approached them together with Penniston. He says he is unable to remember seeing any actual craft and therefore he cannot vouch he saw a craft takeoff. Penniston claims, I believe, that Burroughs did see that and Burrough's sketches of the craft made at the time confirm that, as he later testified during his debriefing in the AFOSI building. Penniston also implies that Burroughs witnessed the takeoff that he described above. That makes two witnesses, not eighty two,—and just one, Jim Penniston, who claims that description of the takeoff. Even Airman Cabansag, who could not have been too far away from the clearing, does not claim that he saw a craft take off.

If the UFO was a real physical craft weighing about 4 tons I suggest that this description of the takeoff is pure fiction:

> "After 45 minutes the light from the craft began to intensify. Burroughs and I then took a defensive position away from the craft as it lifted off the ground without any noise or air disturbance. It maneuvered through the trees and shot off at an unbelievable rate of speed. It was gone in the blink of an eye".

Sound like how UFOs fly in Hollywood movies? Ewoks in the Star Wars movies zipped through the forest trees on flying Speeder Bikes. From a hovering start Han Solo and Chewbacca in their Millennium Falcon spaceship zoom away into hyperspace at "speed—impossible". Did that really ever happen in Rendlesham

Forest? Of course not! That's a fabrication and a fantasy presumably to convince UFO fans that the Rendlesham UFO was indeed the real McCoy. Alternatively Penniston's memory of the UFO's lift off and departure could have been something that was implanted in his memory by his NSA/DIA interrogators using hypnosis. So let's get back to the real world and consider how the UFO might actually have departed from the forest!

If Burroughs and Penniston's real physical UFO was a craft weighing about four tons it would require a powerful propulsion system with thrust in excess of four tons to lift it vertically from the ground. The AV-8 Harrier Jump Jet whose jet engine gives it such vertical takeoff capability produces, I can assure you, a totally deafening roar if you are anywhere nearby when one lifts off. If this UFO was anything like the Apollo Lunar Module (LM, not CM) it would need a vertically-mounted rocket engine of similar thrust to lift it from the ground.

The Apollo CM, boilerplate or otherwise, has no propulsion

The 67th ARRS (Aerospace Rescue and Recovery Squadron) was based at RAF Woodbridge from 1970 on. During the Apollo Program years which ended in 1972 67th ARRS practised recovery operations (right center). Below boilerplate Apollo CM-1206 at RAF Woodbridge.

Below: Folly House at end of RAF Woodbridge runway

Above: Boilerplate CM BP-1206 was used by 67th ARRS at RAF Woodbridge before 1977, sometimes flown slung under a helicopter. It was seen like that by a forester at Folly House late on 25th December 1980.

Above: Sikorsky HH-53 Super Jolly Green Giant helicopters and later MH-53 Pave Low Special Ops helicopters were those used by the 67th ARRS based at RAF Woodbridge in 1980. Note the absence of identification markings here.

system of its own apart from the stages of the huge Saturn rocket which initially lifted it into earth orbit and the CSM rocket motor which propelled it on towards the moon. So how could a BP CM possibly have possibly arrived and departed from that clearing in Rendlesham Forest?

ACTUAL WAY THE UFO ARRIVED AND WAS LATER REMOVED

If the Rendlesham UFO was in fact an adapted Apollo BP CM it must have been taken there by one of 67th ARRS's helicopters. It must have been slung under the helicopter and lowered into the small forest clearing under cover of darkness. There is every reason to think that the selected position, one mile east of RAF Woodbridge's East Gate, was carefully chosen in advance and also the date on which to do this.

The specially adapted BP CM would probably have been flown direct from the US to RAF Woodbridge where the 67th ARRS was stationed. It may have arrived there several days before the Rendlesham Forest Incident. With occasional transatlantic military supply flights into the twin bases during the 1970s/1980s arrival of a C-141 transporter plane would probably not attract much attention. Its special cargo concealed under a tarp could easily been have been put in a hangar out of sight until the time it was needed. The pilots of the 67th ARRS—who were training as experts at infiltration and exfiltration—would already be familiar with ferrying slung loads below helicopters and were among the USAF's best at flying such low-level missions.

If this scenario is correct, a small Special Ops team would have flown with the BP CM from the USA. They would then be there to operate the THW from a concealed nearby position in the forest. First a 67th ARRS helicopter would have briefly hovered over the BP CM on some part of the airfield at RAF Woodbridge while a suspension cable was attached. Then it would have flown east with its large conical slung load at a fairly low level along the main Woodbridge runway, past the forester's house by the end there and towards the forest clearing.

Once the CM had been lowered into position, sitting on its three-legged support frame in the clearing, the helicopter would have withdrawn, perhaps circling back round to some outlying part of the airbase. No doubt the 67th ARRS helicopter would have remained in radio contact with the Special Ops men in the forest using a secure channel inaccessible to any outside party.

Two or three men, possibly from the 160th SOAR(A) Night Stalkers, would have stayed near the CM in a concealed position to operate it. That might have been done by using radio signals or else via a hidden cable connecting a hand-held control panel to the CM. The various lights plus a possible DEW attached to the CM would have been controlled like this, say, by two men hidden in the forest nearby. Besides the bright lights on the CM, one critical element of this THW's operation would have been a small radio transmitter on the CM with which to jam or interfere with radio communications between the base security guards in the forest and the CSC.

So when the time came to remove the BP CM from the clearing how could that have been achieved without it being noticed? I suggest in exactly the same way as it got there in the first place. If the three base security guards, Penniston, Burroughs and Cabansag, spent, let's say, 20-30 minutes walking further to the east before they returned towards RAF Woodbridge it would have certainly allowed plenty of time for the 67th ARRS helicopter to remove the BP CM from the clearing in the forest.

The Special Forces men would have known what position Burroughs, Penniston and Cabansag had reached from their attempted radio messages back to CSC. From the beginning they would have known what frequency the guards' radio communications used and they probably jammed or blocked this when the men first approached the BP CM in the forest clearing. They would most likely have had night vision scopes to watch the three base security guards after they had come out of the forest and were walking through open fields.

When the coast was clear the 67th ARRS helicopter most likely flew in at low level and the Special Ops men in the forest helped attach the BP CM to a cable by which it was winched up above treetop-level and flown away. The exfiltration operation probably took not much more than 5 minutes for the experienced pilots. The HH-53's engine noise was likely have been so low, given the use of LOT, as to go almost completely unnoticed except by someone in the immediate vicinity.

If an Apollo CM was to be used in this manner as a THW in a live anti-terrorist situation some might wonder whether the Special Ops men operating it would have been better off doing that from inside the capsule. After all, the CM's stainless steel shell would presumably be strong enough to protect them from rifle fire should approaching terrorists decide this was a trap and try to riddle the capsule with gunfire. In a live situation that might have been a preferred option. With boilerplate BP-1206 that was not even possible since it had no usable crew hatch to give access to the interior. Also, if there was a man inside the CM, he would definitely have needed a periscope to see anything outside through the docking tunnel at the top.

IF UFO WAS A THW, WHAT HAPPENED WHEN THE MEN APPROACHED?

We are told that when Penniston and Burroughs approached the craft in the brightly illuminated clearing there was "a silent explosion of light." Penniston says they instinctively hit the ground—or else the bright flash may have physically thrown them both to the ground. Could this flash from the top of the craft have been a burst of electromagnetic energy from a Directed-energy Weapon (DEW) activated by a concealed Special Ops man nearby?

It certainly seems to have affected Burroughs and after that flash one gets the impression from what Penniston has written that his colleague seemed to start acting in a concussed, confused, perhaps zombie-like, manner. Burroughs's recollection from that point on appears to exclude any mention of seeing an actual physical craft there in the clearing—despite the fact that Penniston claims both of them watched the alleged takeoff of the UFO.

If Burroughs had been zapped by a DEW, the same doesn't seem to have applied to Jim Penniston. It might have been because he wasn't in the line of fire or else the zapping could have had a far lesser effect on him. He says "We completed a thorough on-site investigation, including a full physical examination of the craft." This apparently included notebook entries and sketches of the symbols, photographs, and also radio relays through Airman Cabansag (who was supposedly at least 50 yards away outside the clearing). All of this allegedly close inspection of the landed UFO took, according to Penniston, about 45 minutes before he and Burroughs—and/or the UFO—departed from the clearing. It was also during this time that Penniston claims he received from the craft a long "download" of binary digits to his brain which he was able to remember perfectly and record in his police notebook more than 24 hours later.

There is no reason to think that any of the reported encounter is fabricated but Penniston's account does beg the question of what Burroughs was doing during those 45 minutes. Was he fully conscious and right alongside Penniston as he photographed and sketched the craft and its symbols as that "we" implies? Did Burroughs also touch the UFO and its symbols? Did Burroughs also receive a download of binary digits that he was unable to recall?

Or was Burroughs lying unconscious somewhere in the background? Had he wandered off out of the clearing in a concussed state? A later version of the UFO encounter in the clearing is given by Penniston in the ERF book he co-authored. He says:

> *I hear a movement to my right; it's John. I am wondering where he had come from and where he was during this encounter. My thoughts are broken with John quickly pointing to a direction out into the farmer's field. "There it is, Jim, let's go!" As quickly as he said that, he was off and I was right behind him.*

So, Burroughs *wasn't* with Penniston all of those "45 minutes" when he says he was examining the landed UFO? No wonder he has no recollection of any craft sitting there in the clearing, only lights. He suddenly re-appears and leads the way out of the forest towards the farmer's field with Penniston following. How very curious that the UFO had supposedly vanished away in a flash and the blink of an eye!

Another way of looking at it is that the UFO simply didn't depart at that point. It may have switched off all its lights and the two men left the clearing assuming it had vanished. Or the two men, both terrified by their shocking encounter with the unknown object, may have fled much sooner than Jim Penniston admits. It would certainly explain his supposed 45 minutes of "missing time" that otherwise can't be accounted for. The two men then followed a light in the sky further away to the east. From the various accounts—sanitized and otherwise—it seems they joined Cabansag and proceeded with him on foot a long way to the east, maybe as much as a further two miles.

Eventually the three men realized that what they were chasing was the revolving beacon light of the Orfordness lighthouse that was four miles further east. The lighthouse in 1980 had a yellowish-white beam which swept around regularly every 5 seconds. Direct view of the lighthouse would only be possible from only a few places at the edge of the forest but its beam would have been scattered by any mist or low cloud that was in the night sky. The yellow light of this beam may have been visible from various positions which they passed when going out into the forest and it would have been in the same direction as the red, white and blue lights of the landed UFO. Nevertheless all of the witnesses are quite adamant that the lighthouse beam never did account for the lights which they saw in among the forest trees.

If the three men spent an hour or more walking a long way further east and then back towards RAF Woodbridge it would have allowed plenty of time for the removal of the craft from the clearing in the forest if it was indeed a BP CM. The Special Forces men would presumably have known where Burroughs, Penniston and Cabansag were from their attempts to radio messages back to CSC. They would certainly have known what frequency the guards' radios used and may well have been jamming them when the men first approached the BP CM in the clearing. When the coast was clear the 67th ARRS helicopter probably flew in at low level and winched up the BP CM.

One additional uncertainty in the witness accounts is the reported noises which were heard near or in the farmer's field at this time. Burroughs says there were strange noises like a woman screaming. It's also suggested there were farm animals in a frenzy there making a lot of noise. Nick Pope suggests the noises could have been muntjac deer barking in the forest. According to Burroughs there were lights on in the farmer's house by the field which they passed through. Did anyone ever go to ask the farmer whether he had heard or seen anything the next day? Or if there had in fact been cattle there at the farm? Was anyone in the house? Apparently these questions, whether relevant or not, were never asked or answered. Of course one possibility could be that the Special Ops unit used recorded sounds as a distraction or to mask their furtive night time activities.

There is some confusion too over the timeline for these events out in the forest. In Penniston's recollection he, Burroughs, and Cabansag set off into the forest to investigate the strange lights that had been seen there shortly after midnight (0000, Dec 25/26). If that is correct Penniston and Burroughs should certainly have reached the clearing where the craft was one mile away from East Gate well inside an hour even if they had walked the whole way on foot.

Burroughs in his initial report—and also Lieutenant Fred Buran who was at CSC at the time—indicate that the men set off into the forest at about 0300 or soon after. Buran also says specifically that he gave the order at 0343 for the men to return to RAF Woodbridge and resume their normal duties. If that time is correct one could assume they all got back to East Gate by about 0430. But that hardly allows any time for Penniston's supposed 45 minute inspection of the landed craft in the clearing or for a further estimated hour for the three men to walk an additional mile or more east beyond the farmhouse and back from there.

When Penniston and Burroughs were summoned back to base (at supposedly 0343) one or both of them retraced their way back through the clearing where the UFO had been. Sure enough there, on hard ground in the very center of the clearing, were three distinct indentations forming an exact equilateral triangle. It seemed plain an object weighing several tons had rested there on its three tripod legs.

Here was definite proof that a real physical craft had recently been there in the clearing and now it was gone. Obviously, they may have thought, it must have taken off under its own power and shot away into the sky as only a UFO could. Perhaps the men convinced themselves they had seen that happen and it became part of their memory of what they had experienced earlier in the night.

INTERROGATION OF PRIMARY WITNESSES BY AFOSI & CIA

On the morning of December 29th 1980, little more than 24 hours after Colonel Halt had returned from his investigation in Rendlesham Forest and from that dramatic UFO display which he and his group had seen, the serious debriefing of witnesses began. This was carried out in the Air Force Office of Special Operations (AFOSI) building at RAF Bentwaters.

Normally any such important AFOSI investigation would be carried out by the senior local AFOSI Special Agent and other local AFOSI Special Agents. Apparently Special Agent Wayne Persinger was deputy commander of AFOSI at Bentwaters at the time when the RFI occurred in December 1980. However Persinger was away on leave at the time of the incident and was not involved in the investigation when he returned. His AFOSI boss, Chuck Matthews, wasn't even stationed at Bentwaters and years later there was disagreement as to whom Persinger said was his boss and the man Bentwaters Wing Commander Gordon Williams remembers as Persinger's boss.

In any case two American Special Agents dressed in plain-clothes conducted the first of the individual debriefings in Major Zickler's office. These gentlemen had apparently flown in from Washington and, according to Burroughs and Penniston, they were from the National Security Agency (NSA) and/or the Defense Intelligence Agency (DIA). Also involved in the interrogations, though in their own separate offices, were Base Commander Colonel Conrad and Deputy Base Commander Colonel Halt.

Statements were taken and drawings were made. Penniston made a four page handwritten statement for the agents which he dated and signed. He was instructed that an official investigation was taking place and that, if asked about the incident, he was to give the cover story provided in a typed statement that was then issued to him. This cover story omitted many of the details he had given them and stated that he had never got closer than 50 yards from the object which they had found in the forest. Burroughs and Cabansag also provided handwritten statements about their involvement in the UFO encounter and later received brief sanitized versions of these which stated only as much as the men would be allowed to admit.

All the direct witnesses were told to treat discussion of the incident in the forest as Top Secret and it is plain that the interrogators intended the whole matter should be contained right then and never allowed to see the light of day. In retrospect, it certainly makes it look much more likely that RFI was a secret military exercise rather than some random excursion by extraterrestrials who happened to land their UFO in Rendlesham Forest right next to a major military airbase!

After debriefing by the agents who had flown in from Washington, Penniston and Burroughs were both further questioned by Base Commander Colonel Conrad and then by Deputy Base Commander Colonel Halt. Further statements were written and drawings were made. Then, together with Cabansag, they were taken to Wing Commander Gordon Williams' office where both Colonel Conrad and Colonel Halt were also present. The sanitized accounts prepared by the NSA agents were presented to the officers and it was again made clear that, since this was a classified matter, the names of the men involved would never be made public. As the senior officer Gordon Williams thanked the men for doing their job and said their reports were appreciated. But, unbelievably, he never asked them any further questions. Not a word, not a single inquiry about their experiences.

The only conclusion is that Colonel Williams must by then

have been made aware that what had happened to the men in the forest had been a secret military exercise. Even if he had no inkling of that when it all began on December 25th/26th it does seem that by December 29th he must have been told it was a Special Ops test using a secret weapon and that fact should never be disclosed. Extraterrestrial aliens in their UFOs? I hardly think so!

There are five of these somewhat sanitized statements produced by the NSA after debriefing five of the witnesses that have entered the public domain. Besides the statements attributed to Penniston, Burroughs and Cabansag, there are statements which are attributed to Lieutenant Fred Buran and MSgt J.D. Chandler. None of the statements use the term "UFO" but besides telling of lights in the forest and possible lights in the sky, there is reference to an object "that was definitely mechanical in nature" by Jim Penniston. That was certainly sensational but there's a lot more that is left unsaid.

There is no indication that Colonel Halt himself was interrogated by the agents who came to Bentwaters to quiz those men who were involved on the first night of the RFI. No doubt he briefed his superior officers at Bentwaters after he had gone out on the final night of activity but there's certainly no indication that he was ever let in on what Colonel Williams must by then have been told. Apart, of course, from the fact this was all Top Secret.

Another very significant thing about the interrogations in the AFOSI building is that no debriefing ever seems to have been done with those who had gone out with Halt on the night of December 27th/28th. Lieutenant Bruce Englund, MSgt Bobby Ball and Sgt Monroe Nevels had accompanied Halt into the forest yet it seems the intelligence agents weren't interested in what they might have had to say about what they had seen. Apart from them, we are told there were several other men in the forest who must have seen the supposed UFOs that were witnessed by Halt. No statements were taken from any of them so far as we know.

Is this because the THW which I suggest was placed in the forest on the first night did not make a second appearance there on the night of December 27th/28th and it was the effectiveness or otherwise of this THW that the NSA agents from Washington were *really* interested in. Not the "UFO flying display" using lasers and LPIs that seems to have been put on for Halt and his men on the night of Dec 27th/28th! This conclusion is very much supported by the fact that Burroughs and Penniston both say they were treated like enemy combatants in the harsh interrogations and those which followed. If such a THW was soon to be used against an enemy, as it might well have been in Iran, those who had devised it must have been determined to know if it was going to succeed in its proposed Honey Badger mission.

Penniston says that he went through at least fourteen debriefings including two that were by non-USAF personnel. He says he was methodically and consistently interviewed and interrogated by his chain of command and other agencies. Every time he was promised that this was the last interview and it would be absorbed into the classified annals of data and he would need not to talk about it any more. This, of course, was not the case.

The debriefs were all "for the last time" he was promised. Tell all and tell it correctly and it would be the last of questions about RFI. But unfortunately these were to continue no matter what he said.

This does makes one think that when, in 2010, Jim Penniston first told anybody about the download of a binary code into his head—which he received directly from the UFO in the forest—he may well have been telling the truth after all. The long sequence of '1' and '0' binary digits was recorded by him in his police notebook after he returned to his digs in Ipswich. That part of his story is certainly controversial and it does seem doubtful the code sequence contains any message of significance. We will look at that later when considering the possibility that an experimental THW might have been designed to imprint some basic form of identifier or potential mind control instruction on a person who approached it or touched it. Not that such a code would necessarily enable any sort of mind control over a person but the method could show if such imprinting was even possible. It could also, as I suggest, have been a novel means of psychotronically tagging a person for future identification in a way that one might wish to tag a terrorist.

Amazingly Penniston never told his interrogators about his police notebook which he kept locked away in his room in Ipswich. Besides the binary code sequence, he had used this notebook for his sketches of the UFO and also of the symbols and glyphs that he had found on its surface. He says he gave them all the information and drawings that they wanted from memory and at no time was the notebook ever mentioned. Despite the fact that sodium pentothal— the so-called "truth serum"—was apparently administered several times by his inquisitors, presumably he never yielded up the secret they were looking for. Surely what they were after must have been the binary sequence which the agents knew had been implanted in his brain? However he seems to have had no idea what the ones and zeroes were back in 1980, let alone they represented a binary code. If the concept and the significance weren't something he'd given any thought to, it's unlikely even sodium pentothal could drag the sequence out of him!

Although some officers who were based at Bentwaters/ Woodbridge at the time have since tried to downplay the brutal interrogation which the primary witnesses to the first night's events underwent, Colonel Halt leaves us in no doubt as to the severity of it. Writing in Leslie Kean's book *UFOs: Generals, Pilots and Government Officials go on the Record*, Halt makes the following extraordinary claim:

> *"Air Force O.S.I. (Office of Special Investigations) operatives harshly interrogated five young airmen, some of them in shock at the time, who were key witnesses. Drugs such as sodium pentothal, often called a truth serum when used with some form of brainwashing or hypnosis, were administered during these interrogations, and the whole thing*

had damaging, and lasting, effects on the men involved."

One doubts that Colonel Williams was aware of all this at the time and of any dire psychological or physiological effects experienced by Penniston and Burroughs as a result of their encounter and the harsh interrogations. It's quite possible he was given a plausible version of events that omitted much of the detail and also the true purpose of the exercise that had been carried out by a small unit of Special Ops men in Rendlesham Forest. Years later he told Burroughs and Penniston he knew they hadn't been treated fairly and that AFOSI's Wayne Persinger "could be real tough". That was very surprising since Persinger hadn't even been involved in the interrogations which were initially carried out by two NSA/DIA agents from Washington.

Likewise it would seem Base Commander Colonel Conrad might not have been told the whole story at the time. However he was obviously aware at an early stage that the whole business was considered Top Secret and that it must be contained. He it was who confiscated the blotters containing the first reports and it seems very likely that he ordered confiscation of all the film taken of the UFO and markings made by it in the forest which had been sent to the base photo lab.

Conrad must have realized that the truth about the RFI was one that could never be made public and in August 2011 he tried to present a totally skeptical view of it when talking to the UK's *Daily Telegraph*. He said "We saw nothing that resembled Colonel Halt's description either in the sky or on the ground." This very dismissive line implied that Halt was unreliable at best or else a fantasist at worst. To me it sounds like the fury of a man who knew the whole affair had to be contained but this had all been thwarted by the FOI release of Halt's memo. For that fiasco he evidently still squarely blamed the unfortunate Halt even 30+ years after the RFI events.

On the night of December 27th/28th when Halt was sent out into the forest by Conrad, the latter went back to his home on the base after the award-giving dinner. For some of the time he was in radio contact with Halt and when Halt and his men reported (from near Green Farm) they were seeing UFOs in the sky, Conrad went outside with his wife to look. Other senior officers at the base did the same. Despite the fact that it was a clear cloudless night he says they saw nothing suspicious. He said there was no hard evidence of anything unusual going on and remains insistent that the radiation readings Halt found using an APN-27 Geiger counter at the supposed UFO landing site in the forest were nothing other than "normal" levels of background radiation.

How then does Conrad explain the prolonged and harsh interrogation of Jim Penniston and the other men by the AFOSI Special Agents (or perhaps agents from NSA and/or DIA)? As one might expect, Ted Conrad claimed that these debriefings have been highly exaggerated and sensationalized.

Of course Charles Halt was having none of this and he fired a letter back to the newspaper as follows:

> *Ted Conrad is either having memory problems, has his head in the sand, or is continuing the cover-up . . . Thru the years Conrad has made conflicting statements about the events . . . . Now he is smearing those involved. It's pretty clear there was a very intense confrontation with something in the forest. Does Conrad want to talk about how the airmen were then subjected to mind control efforts using drugs and hypnosis by British and American authorities?*

Who is one to believe? I suggest that former Base Commander Ted Conrad is one of the few people at the base during that week in December 1980 who knew near enough what was happening and that it was Special Ops men who were responsible for the RFI. However, Conrad's military oath of secrecy will probably never let him reveal it.

Part IV

DEWs and Psychotronic Weapons

Directed-energy weapons (DEWs)—or beam weapons—appeared in popular fiction as early as the turn of the 19th century. A "heat-ray" is described by H.G. Wells in his 1898 novel *The War of the Worlds*. Science fiction over the next thirty years often creatively depicted a number of similar weapons such as the ray gun, the death ray, the disintegrator ray, blasters, and plasma rifles. Most memorable of these ray guns were the *Star Wars* "phasers"—which seemed to be the weapons of choice a long time ago in a galaxy far, far away.

Many of those fictional DEWs—popularized well before the term DEW was invented—appeared to be fiendish weapons of unimaginable destructive power. Of course, some military commanders in the first half of the 20th century may have wondered whether such weaponry might one day be available to them and give them the edge over their enemies in war. Not only for use against enemy combatants but maybe there could be a ray gun which would shoot down enemy planes and missiles?

If beam weapons like this could be developed and used to destroy offensive aircraft and missiles it was clear that very high energy sources would have to be directed and focused in order to achieve such destruction. Unlike the shells and bombs that had been used by the military of both sides in World War I and World War II beam weapons would require entirely new scientific techniques. With the advent of atomic energy and nuclear bombs at the end of World War

II such weapons now looked to be in the realm of possibility.

As described earlier, US General George Keegan had claimed in the late 1970s that the Soviets were testing a particle beam weapon which, if successful, would be capable of destroying US ICBMs fired at targets in the USSR. American reconnaissance satellites had photographed some kind of DEW installation at the main Soviet atomic bomb test site near Semipalatinsk in central Asia. The energy source for this DEW was said to be an underground steel sphere in which small nuclear devices were to be detonated so as to allow beams of protons or other ionized particles to be directed at targets by using powerful magnets. Such beams would be capable of destroying incoming ballistic missiles or aircraft. Keegan urged Congress to fund a similar particle beam program or else the US nuclear deterrent would become worthless. That would allow the Soviets to embark on a first nuclear strike against the US with impunity. He warned that without suitable US investment a beam weapon gap might open up giving the Russians a dangerous advantage.

Keegan's plea was mostly met with skepticism and a reluctance to extend the arms race with the Soviets even further. Later evaluation of intelligence showed the particle beam weapon being tested by the Russians had little if any success and it was eventually abandoned. Nevertheless his warning was sufficient to produce some support in the US government for the Strategic Defense Initiative (SDI) which was announced by President Reagan in 1983. The proposed SDI missile defense system –derisively known by some as the "Star Wars Project"—was intended to protect the US against ICBMs and submarine-launched ballistic missiles.

Although most of SDI's "Star Wars" technologies never reached fruition during the 1980s, research into various types of DEWs continued even after the end of the Cold War. Particle beam weapons which would probably have needed large nuclear linear accelerators or else cyclotrons being launched into Earth orbit were forgotten but development of other smaller DEWs such as lasers

Above: A directed-energy weapon that uses lasers, microwaves or particle beams against a target.

continued in the US. As we saw earlier, much of the development of military lasers was carried out at Kirtland AFB, New Mexico.

Donald Rumsfeld, who was US Secretary of Defense under President George W Bush, reputedly once remarked that the purpose of war was to kill the enemy. Although this was sometime after the deadly 9/11 terrorist attacks by al Qaeda in 2001 and such a sentiment had wide popular support in the US, there were many critics who disputed such belligerence. There were alternatives to killing enemy combatants available and such techniques were known as "non-lethal warfare".

There were, of course, non-lethal weapons used by police rather than the military well before anyone considered whether some kind of DEW might provide a solution to the requirement. Police truncheons, water cannon, tear gas, pepper spray, rubber bullets and tasers spring to mind but even these aren't always non-lethal. Maybe the term "less lethal" would be far more appropriate? Such weapons were primarily needed to control rioters and disperse angry mobs on city streets.

Apart from massive particle beam weapons and powerful lasers that were intended to destroy enemy missiles and aircraft, some smaller scale DEWs were also being developed as non-lethal weapons. One instance of that was the Active Denial System (ADS) or heat ray that was deployed by the US military in Afghanistan in 2010. ADS fires a high-powered (100 kW) beam of 95 GHz electromagnetic waves at a target and works on a similar principle to a microwave oven. The intention is to cause severe pain by heating the water in a human target's skin and its principal function was for riot-control duty. Although ADS supposedly caused no lasting damage to those it was aimed at, there was concern it could cause permanent damage to the eyes. It had various sizes, including one that fitted onto a Humvee.

A different variety of supposedly non-lethal DEWs pioneered in the 1970s and 1980s was one which was termed a "psychotronic" weapon. These were electronic devices which directed beams of non-ionizing electromagnetic energy directly at people's brains and/or bodies through the air. Particular EM frequencies could apparently produce results akin to brainwashing or even rudimentary mind control. It was then suggested that a target person's memory of events could be erased or altered by such a weapon. There might also be temporary physiological effects such as migraine and possible tissue damage to an organ like the heart. Strong RF fields in the immediate vicinity of powerful radio transmitters are known to cause headaches and have other bad effects. They can also interfere with cardiac pacemakers.

The first time I heard the term "psychotronic" was at a conference on anomalous phenomena in Atlanta, GA, in April 1992. It was at the fourth meeting of TREAT (Treatment and Research of Experienced Anomalous Trauma), a group founded by New York psychiatrist Rima Laibow to deal with PTSD in patients who believed they had been abducted by ET aliens. That, together with UFOs, remote viewing and, of course, psychokinetic spoonbending and the like, seemed to be the main theme of this particular TREAT conference.

Psychotronic warfare and, by implication, generation of EM beams with psychotronic effects on people was one subject in a lecture by Colonel John B Alexander. I'd no idea of what the word psychotronic meant and how it was in any way connected with the UFO subject. Evidently such an electromagnetic beam could, in theory, induce altered states of consciousness and/or hallucinations in a targeted person—whether they were aware of being targeted or not. If one such altered state was that of supposed abduction by aliens, rather than any physical reality, would a targeted person remember it later and could that be recalled by using hypnotic regression? Who would use such weapons against unwitting victims? Were psychotronics something that supposed aliens had access to—or was this just a theoretical weapon that might be being developed by the US military? We were not told if such weapons actually existed or whether they had ever been tested on people.

John Alexander had served for many years in the US Army and was Commander of Army Special Forces Teams (Green Berets) in Thailand and Vietnam during the 1960s. He was best known as a leading advocate for the development of non-lethal weapons and military applications of the paranormal, known then as "psychic warfare". He also wrote and lectured on the reality of UFOs. When working for the DIA in 1985 he founded what was called the ATP (Advanced Theoretical Physics Project) in order to discover whether there was a secret federal government UFO project. His group consisted only of those who had a Top Secret security clearance—and since the group's name made no mention of "UFO"—its purpose long remained unclear to outsiders. Although Alexander says his group never found any evidence that the US government ran a secret UFO project, he still firmly asserts his belief in the reality of UFOs.

He retired from the Army in 1988 and not long after was appointed Program Manager for the Non-lethal Defense Project at Los Alamos National Laboratories (LANL) and lived near there in Santa Fe, New Mexico. Whether or not his Non-lethal Defense Project at Los

Above: Retired US Army Col. John B. Alexander, who in 1985 founded the Advanced Theoretical Physics Project to find out if there was a secret federal UFO project. He was later Project Manager for Non-Lethal Defense at LANL.

Alamos included the developing of psychotronic weapons we can only speculate. The Atlanta TREAT conference took place more than ten years after RFI and there was certainly no mention of that at the time. If Alexander knew that RFI had involved a psychotronic weapon, he was keeping mum.

A psychotronic DEW is a weapon that can be used discreetly, often without a targeted person even knowing anything about it. If most of the EM radiation produced by a DEW is of a frequency above or below the visible spectrum there would be no flash and no sound when the weapon was discharged.

If the supposed UFO in Rendlesham Forest in December 1980 was equipped with a DEW that was fired at Burroughs and/or Penniston that would certainly explain the flash which caused them to throw themselves, or be thrown, to the ground. It would also imply that the principal EM frequency of its discharge pulse was in the visible spectrum since they were immediately aware of it. Even so, a more powerful part of the EM pulse might have been of different wavelength to visible light—in, say, the invisible submillimeter or millimeter range of microwave radiation.

Although DEWs may be produced that deliver different EM frequency ranges, microwave energy with frequencies of between 1GHz and 100 GHz are generally believed to be the most dangerous when aimed at humans. Of course the intensity—or EM energy flux—of such a beam and the effect on a person will depend on the input power of the DEW and also how near the (human) target is. A targeted person standing, say, within 100 ft (30 m) of such a weapon could well be severely affected.

Common physiological effects of non-lethal electromagnetic DEWs include:

- Disorientation
- Difficulty in breathing
- Nausea

- Blurred vision
- Pain, systemic discomfort
- Loss of balance
- Vertigo
- Memory loss

Pain and discomfort caused by a non-lethal DEW attack would likely confuse, disable or perhaps subdue a victim who was directly zapped. It might also cause temporary or permanent loss of immediate memory. Longer term effects of sustained DEW attacks on human targets may include permanent brain damage which is something that has become more apparent during recent years. Back in 1980 John Burroughs appears to have been badly affected as regards memory loss during his encounter with the UFO in Rendlesham Forest.

Directed-energy weapons that target the central nervous system and cause neurophysiological disorders may violate the CCWC (Certain Conventional Weapons Convention of 1980). In going beyond non-lethal intentions and causing "superfluous injury or unnecessary suffering" they may also violate Protocol 1 of the Geneva Conventions of 1977.

If the US military tested such weapons on human guinea pigs during the 1960s and the 1970s it seems likely that any soldiers or civilians who were targeted were quite unaware of it and were certainly were not volunteers. Numerous unethical human experiments are known to have been performed on test subjects during the second part of the 1900s without people's knowledge, their consent or even their informed consent. The secret experiments included administering of untested drugs, radioactive substances, hallucinogens and other mind-altering substances. Public outrage in the late 20th century over disclosure of such government experiments on human subjects led to numerous congressional investigations and hearings.

However it wasn't just the US military and intelligence services who seemed to be testing such potential weapons in the 1960s and 1970s. The so-called Moscow Signal was a reported microwave transmission varying between 2.5 and 4 GHz that was directed at the US Embassy in Moscow from about 1953 until 1976. The directed energy beam was intermittent and of low power (under 5 microwatts per sq cm) and came from a Soviet apartment building about 330 ft(100 m) from the ten floor embassy. US authorities who detected and monitored the microwaves kept details of it secret from the US public and even from the embassy staff who had been working there at the time. The RF signal put out by the Soviets had less power than a microwave oven but, nevertheless, a hundred times more than Russia's maximum RF exposure standards which caused concern among US officials. It seems the effect of the Moscow Signal on staff at the embassy was mainly due to long term exposure rather than short RF pulses.

Tests on various members of staff who had been exposed to the Moscow Signal showed significant lymphocyte abnormalities in the blood and also an elevated rate of cancer among personnel. No shielding against this DEW was put in place for 11 years after the signal was first detected and there was also great controversy whether the Soviet intention was to degrade the health of the US Embassy employees or whether it was really intended to activate electronic eavesdropping devices which the Russians had secretly installed in the building. The long-running Moscow signal episode certainly led to great paranoia, justified or not, during the Cold War years.

At the time of the Moscow Signal there was no suggestion of mind control and the word psychotronic was yet to be used in that context. In any case there can be little doubt that development of psychotronic weapons continued secretly on both sides of the Iron Curtain. However, from roughly the time of that TREAT conference in Atlanta, the word "psychotronic" was seldom used in scientific literature and even now it does not appear in most dictionaries. In one the word is just defined as an adjective relating to a genre of movie film which is characterized by bizarre or shocking story lines. Not that there was a conspiracy of silence as regards this word but if such psychotronic research did continued the US, it was probably described differently!

From the 1950s through to 1973 the CIA ran a secret mind control program called MKUltra. It was intended to manipulate people's mental states and alter brain function, including surreptitious administration of drugs (especially LSD) and other chemicals, sensory deprivation, isolation and various forms of torture. The human guinea pigs used in this testing were often unaware of what was being done to them without their knowledge or their consent. Some were psychologically or physically damaged and some died.

MKUltra was organized through the office of Scientific Intelligence of the CIA and coordinated with the U.S. Army Biological Warfare Laboratories. The operation supposedly ended in 1973 but there is still controversy over whether it really ended or whether it continues to the present day albeit under a different name and under a different branch of the intelligence services.

The direct control of a subject via radio signals aimed at the brain was an elusive goal for those running the MKUltra program and it is very doubtful if any actual mind control was ever achieved. In more recent years scientists have succeeded in controlling laboratory animals such as rats via radio signals to the animal's brain which are picked up by tiny sub-cortical implanted electrodes. Remotely controlled animals like these can be considered cyborgs but the electrodes do not move the animal directly as if one was remotely controlling a robot. The radio-waves were used to signal a direction or action required by the human operator and they stimulated the animal's reward center if the animal complied. Very soon a lab rat may appear to behave like a robot but it is not a true robot. No doubt scientists who have engaged in such mind control attempts similar to MKUltra must have considered whether such techniques could be used to control human beings.

Some rudimentary mind control techniques that were learnt from MKUltra testing might have been available in 1980 but these would have been of little practical use. If DEWs with submillimeter or microwave frequencies were to be used, such weapons could probably be supplied for inclusion in a hastily assembled THW back then. However, if attempted mind control techniques from earlier years had proved unsuccessful, any psychotronic DEWs available then would probably only be capable of stunning and subduing an approaching victim. The point of using a weapon like this would presumably be so the target was unaware of what had happened to him and who was responsible.

Nevertheless the development of such psychotronic DEWs does seem to have continued in the USSR up to the end of the Cold War and then later still in Vladimir Putin's Russia of today. The UK's *Independent* newspaper reported in May 2021 that since 2016 nearly 50 Americans are believed to have been sickened by an unknown 'directed energy' weapon. Further reports in *The New York Times* put this number of US personnel who have fallen ill at over 130. Those attacked were said to have been suddenly stricken with "Havana Syndrome"—the name now given to a mysterious illness that varied but ranged from severe headaches to ringing in the ears, as well as loss of hearing, memory, and balance.

Sudden onset of the illness was first reported among US diplomats in Havana in 2016 and since then US troops and intelligence officers appear to have been targeted in other parts of the world including some cases on US soil. Reports that have trickled out from the CIA and the Pentagon indicate that they don't believe this is a naturally occurring illness—it's a deliberate act of aggression.

As to what is causing "Havana Syndrome" there is little doubt that it was "directed pulsed RF (radiofrequency) energy" according to the National Academies of Sciences, Engineering, and Medicine in a report on the illness commissioned by the State Department. In a *New York Times* analysis the description of the energy being aimed at targeted people included the words "pulsed" and "directed". This was not energy being randomly dispersed by a cell phone or other device. It was energy being quite specifically aimed at particular people.

The reports leaked from the CIA and other agencies have been vague, citing classified information and the need to avoid making unfounded accusations. However some members of Congress were more blunt about what they think is going on. Most suspect that Russia is behind the DEW attacks but, of course, Russia has denied any responsibility.

In May 2021 Senator Susan Collins, who is on the Senate Intelligence Committee, told CNN "There's a mysterious direct energy weapon that is being used. And it is causing, in some cases, permanent traumatic brain injury." If confirmed, these directed energy attacks would join a long list of non-military assaults Russia has waged against the United States in recent years, including the Solar Winds cyberattack and also interference in the 2016 presidential election.

It certainly looks as if the secretive use of DEWs against specific human targets in various parts of the world continues today—as it has to a greater or lesser extent over the last fifty years. This sort of DEW is very likely the same sort of psychotronic weapon tried out on the unsuspecting John Burroughs and Jim Penniston in Rendlesham Forest, England, in the early hours of December 26th 1980.

HEALTH PROBLEMS AND INJURY SUFFERED BY BURROUGHS

Both men were subject to various physiological and psychological effects following their encounter with the UFO in Rendlesham Forest in the early hours of December 26th 1980. They both appear to have suffered fairly severe post-traumatic stress disorder (PTSD) during the aftermath of RFI. PTSD is an anxiety disorder that follows the trauma of some event which has threatened serious harm or death. Its symptoms include unpleasant memories of the event, nightmares, headaches, intense guilt or worry and life-changing anxiety which can dominate one's existence.

In the case of Jim Penniston some of this PTSD might be put down to his prolonged and ruthless interrogation by AFOSI and/ or other intelligence agents which started a few days after the RFI event. If indeed the truth serum sodium pentothal and other such drugs were used during his interrogation that could have induced psychological disorders.

Within weeks of the incident Penniston was being treated for ailments he had not had previously; headaches, vertigo, memory issues and unknown infections. These were all treatable but they kept recurring. However, during his remaining time in the USAF his memories of what happened in the forest got pushed to the back of his mind. It was only after his retirement that the nightmares and dreams all started to happen again and he began to suffer endlessly from chronic fatigue syndrome.

He says that if his memory was messed with, the question is why? "What was it we saw and why have the United States and the United Kingdom done everything to cover it up as a UFO event?" He asks "What is it they are so afraid of the public finding out?" He goes on to say that whatever it is that they want to be kept so secret obviously has nothing to do with UFOs (ETs or aliens) as is clearly shown in the declassified documents about this phenomenon.

John Burroughs developed serious heart problems from which he very nearly died in July of 2012. As a result he had needed to have a pacemaker defibrillator fitted to control his heart function. His doctor was unable to find the cause of his heart condition and asked that his military medical records be obtained to help him establish the cause. Normally such records can be obtained in the US from the VA (Department of Veterans Affairs) but it soon became evident there was a problem. Burroughs and his attorney, Pat Frascogna, enlisted the help of Arizona Senator Jon Kyl to help obtain the necessary documents. Kyl's first attempt was denied by the US VA who told him Burroughs first needed to file a VA claim for disability.

Then Kyl was informed that the requested medical records were located in the VA's classified records section. It would be up to the USAF to decide whether they could be released to the Senator and to Burroughs' doctor for his treatment—and even for them to decide whether the VA was allowed to treat him. The classified medical records were evidently considered a matter of national security.

John Burroughs says that he was shocked at the time when he realized that the US government could be holding up his treatment because his military medical records were classified. Months went by without him receiving any response on this issue. He sent Senator Kyl another letter which he mailed in November 2012. A month later Kyl wrote back saying that he was retiring and Burroughs needed to try again with Arizona Senator John McCain's office by sending them all of the documentation over again.

During 2013 it does seem that John McCain managed to produce a successful result from the VA and the USAF. John Burroughs was granted his disability claim and further treatment was entirely paid for by the VA. His doctors were presumably informed about the kind of electromagnetic radiation and its intensity to which he had been exposed back in December 1980 but it is not known if all of his classified medical records from that time were released.

Having said that, some information about this radiation has been revealed by Dr Christopher "Kit" Green who headed the CIA's so-called "UFO Desk" back in the 1980s. At a UFO conference in Eureka Springs, AR, in April 2015 the audience were shown his answers to two questions relating to John Burroughs' medical records that had been put to Green—who was not present—in a recent interview:

Q #1 ... Yes, his records were classified. Some remain classified. The reasons are not entirely ethical. Thank god that a couple of Senators had the guts to push, and push to 'Disentangle' that which was legitimately classified from those records that were needed to make the right decision about injury sustained while on Active Duty. Period. It is/was true. It is sensible. It is not entirely the stuff to make us physicians very proud.

Q #2 Broad-band Non-Ionizing Electromagnetic Radiation caused the injuries. The RF is identified in a dozen classified and a half-dozen unclassified studies on cardiological and neurological injuries ... not thousands of reports. Very, very few physicians even care about this arcane area of research ... and fewer know about the injuries sustained by near-field (< 100 M) to humans. The data is sparse. It is not properly Peer-reviewed. It is not understood, it is not the subject of current research. And that is the truth.

"The decision that was made to grant John medical disability was just. Some of his records will remain classified. Those of us in Military and Intelligence Medicine can be proud the right decision was finally, if belatedly made; we should remain both vigilant and ashamed that our profession remains improperly darkened, and we should bring it to the light when we can.

Kit Green added that he had not cleared this short statement with John Burroughs, his Attorney, or anyone in his government including his Contracting Officers' Technical Representatives. The statement is signed with his full name: *Christopher C. Green MD, PhD, FAAFS.*

Since this was being presented at a UFO conference the implication was clearly that Burroughs' injury had resulted from his encounter with the landed UFO in Rendlesham Forest and that the UFO was, presumably, of extraterrestrial origin. To some at the conference the fact that his medical records had been classified by the government implied they knew about the alien threat which must remain secret. Or, worse, the government was somehow in league with the aliens

Not true! In my view Green's statement is clearly proof that the US government recognized its responsibility for having tested a secret weapon on an unsuspecting airman without him being a volunteer.

If his injury was caused by non-ionizing electromagnetic radiation it was almost certainly the result of Burroughs being targeted with a DEW which had been fitted on that boilerplate Apollo CM in the forest clearing. It must have been aimed at one or both of the approaching base security guards by whoever was in control of this DEW. Presumably the bright flash which caused both men to fling themselves to the ground was a bolt of EM energy from the DEW which hit Burroughs but maybe did not directly hit Penniston.

That it may have been aimed at one man could been intentional so that the subsequent differing behaviors of the affected man and his unaffected companion could be evaluated and contrasted. Penniston, who then apparently walked right up to the "UFO" and touched it, was possibly targeted by a quite different device on the CM which he believes downloaded a long series of binary digits directly to his brain and imprinted them in his memory.

PENNISTON'S BINARY CODED "MESSAGE" FROM THE UFO

We now reach one of the most problematic parts of the extraordinary story of Jim Penniston and John Burroughs' encounter with the UFO in Rendlesham Forest. Years after the RFI when Jim Penniston first started to talk about his experiences in public and at some UFO conferences, he revealed the existence of a lengthy series of binary digits, each either 1 or 0, that he said he had written down in his police notebook soon after his 1980 encounter with the landed UFO. It was then immediately suggested that this binary sequence was a special message from the unseen intelligence inside

Above: Jim Penniston and John Burroughs walk in Rendlesham Forest many years after their December 1980 UFO encounter.

the UFO. Some thought ufonauts within must have been ET aliens though Penniston himself claimed they were probably time travelers from the future.

When I first heard of Penniston's Rendlesham binary code message I was extremely skeptical and initially assumed it was simply further embellishment of a highly dubious story. However years later when new information about the encounter was clearly pointing to the fact that RFI may have been the testing of a secret THW by a unit of US Special Ops, I realized his account of "downloading" a binary code was quite possibly true. Let us look closer at this part of his story.

Jim Penniston's first mention of writing down a binary code that he says he received from the Rendlesham UFO was in 1994 some time after he had retired from the USAF. In the book *Encounter in Rendlesham Forest* he describes how, when examining the UFO in the forest clearing and running his hands over the "glyphs" on its surface, he left touching the large central symbol **A** until last. This symbol, he felt, was the key to the craft. He pressed the palm of his hand against this large triangle. Then, no longer seeing what he saw in his mind's eye, he saw a bright and steady brilliant light. A stream of flashing ones and zeros then ran relentlessly and, though scared, he knew it wasn't harmful. No longer able to see his surroundings, he was unable to say how long the bright light lasted, whether seconds or minutes.

Unable to pull his hand back, he says he was eventually released by whatever the device on the craft was. The ones and zeros stopped and he saw his hand was undamaged. Scary and unreal as it all was he knew there was no need to touch the craft again and so he withdrew.

During a hypnosis session he underwent in 1994 it was suggested to him that that the unknown craft was somehow repairing itself and by touching it he had interrupted the process. The hypnotherapist said: "By touching the symbols and you disrupted the repair program?" He replied that he had activated a binary code and the two government agents who were interrogating him wanted to know why.

It was during this hypnosis session, we are told, that the idea the craft is controlled by time-travellers first appeared. Also the description of the ones and zeros as a binary code which was not a term he had used previously. Although it does not appear that he was lead by the hypnotherapist to say these particular things it is just possible that he was hypnotically recalling terms used by his NSA/DIA interrogators years earlier in 1980/1981. It was of course surprising in 1994 that he was now talking of time-travellers, rather than ET aliens, but ever since that time he has publicly asserted he is sure of that. At one point in the session he states "They are time-travellers—they are us." Whether that is what he *really* believes—or not—is anyone's guess. It could just be that was what his NSA/DIA interrogators had told him under hypnosis and the binary code he received could well have been what the government agents were trying to extract from him.

The binary code as recorded by Jim Penniston in his police notebook, just as he says he wrote it, is shown in full in the book *Encounter in Rendlesham Forest,* which he co-authored with John Burroughs and Nick Pope. It is handwritten on the right-hand side pages of the small ring-binder notebook when opened. All left-hand side pages were left blank. The binary digits ("bits") are, of course, always 'o' or '1' with about 16 bits written on each line and about 16 written lines to the page. The lines of bits are written somewhat unevenly and Penniston sometimes leaves blank spaces between blocks of bits. There is no way of knowing whether this long string of binary digits is accurate or how closely it reflects what he claims he telepathically downloaded from the UFO.

The first three lines of the apparent message appears as follows:

01000110010011101010110
101010100100101011
000100100001000011

There follow many more lines of binary written down in his notebook.

For those who are not familiar with binary code it should be said that all modern computers, cell phones, other telecommunications devices and computer peripheral equipment use such streams of binary digits to store data and transmit it wirelessly or otherwise through networks such as the internet. Such digital data can be numeric, alphanumeric, image data, audio data, video data or any other kind. Whatever sort of data is used, it is always encoded as strings of binary digits. Unless the system of encoding that was used to do that is known to someone trying to interpret it, it is impossible to get a meaningful result.

Back in 1980 the most widely used encoding systems available for

US and UK computers, and associated telecommunication systems, were IBM's EBCDIC (or else BCD), and also ASCII. The latter was developed from telegraph code and the FIELDATA system developed as a standard for battlefield information. ASCII was a standard code used for UNIVAC computers and it was also used by several other computer manufacturers.

Of course we don't know whether the supposed ET aliens use streams of binary digits to transmit their data or whether time travelers who supposedly live many centuries from now into Earth's future continue to use digital methods dating from the mid-twentieth century. So how could it be possible to interpret Penniston's alleged coded message?

Evidently a small group of UFO researchers who met with Penniston and Burroughs in Phoenix, AZ, in October 2010 claimed that they could do so. Film-maker and UFO researcher Linda Howe and Kim Sheerin, who was executive producer on the History Channel's *Ancient Aliens* show, were interviewing the two men about the Rendlesham Forest Incident when Penniston produced his original police notebook to check a date. Linda Howe and John Burroughs happened to see the lists of ones and zeros written in the notebook and asked what they were. Rather hesitantly Penniston replied. "Yes, I wrote them down." Everyone present then got very excited about what Penniston had felt for 30 years was a product of his temporary madness resulting from exposure to the craft on that night in 1980.

Linda Howe and Kim Sheerin got their respective "experts" to supposedly decipher parts of Penniston's binary code. Apparently by assuming that the aliens used some variant of ASCII encoding, they later claimed to have extracted translations such as these:

EXPLORATION OF HUMANITY 666 8100
52.0942532N 13.131269W (Hy Brasil)
CONTINUOUS FOR PLANETARY ADVAN???

29.977836N 31.131649E (Great Pyramid of Giza, Egypt)
14.701505S 75.167043W (Nazca Lines in Peru)
EYES OF YOUR EYES
ORIGIN YEAR 8100

. . . . and a few more alleged interpretations which Nick Pope says looks like a New Age wish-list. These appear to include locations of places which might be associated with ETs and ancient aliens!

To draw meaningful numeric and/or alphanumeric information out of a mass of binary ones and zeros by using some arbitrary process similar to the notorious Bible Code is, of course, unbelievable. I suggest such results are entirely worthless. The precise latitudes and longitudes shown for those places which actually do exist can easily

Above: John Burroughs with Nick Pope (co-author of 'Encounter in Rendlesham Forest') and Linda Howe at Ozark UFO conference in Eureka Springs, Arkansas, April 2010.

be found from internet applications like Google Earth. It does seem extremely unlikely that these values for geographic locations, each given to six places of decimals, can really be extracted from Penniston's binary code—even using a home-grown variety of ASCII that the "experts" have chosen. As for Hy Brasil, that is the name of a phantom island from an ancient Irish myth. It was supposedly cloaked in fog except for one day in every seven years. Even then the mythical island could never be reached and obviously it could never have actually had any true latitude and longitude. Hy Brasil is a fantasy place, so the kindest thing one can say about the alleged *Ancient Aliens* interpretations of the binary message supposedly downloaded into Jim Penniston's head is that they are of extremely dubious value, if of any value at all.

So, if like me, you think these interpretations of the binary code are simply contrived messages from the "aliens" adapted for ufotainment purposes by film-makers and TV producers for the History Channel, then could there be any possible genuine meaning in Penniston's binary code? If—and that's a big 'IF'—he really did somehow receive a long string of binary digits from the UFO in the forest clearing, and that he was somehow able to memorize it, and then he was able to write it down in a notebook quite soon afterwards, I can now suggest a rather more plausible explanation.

It was probably not a message being sent to any particular person or even one that any recipient could interpret. The binary data stream may have included coded markers or tags that could somehow be automatically imparted to a person who touched this THW UFO in order to identify that person at a later stage. Any such tags could be in alphanumeric code, say EBCDIC or ASCII, and serve as an invisible identifier label. It would, in theory at least, enable the NSA/DIA interrogators to extract the tag(s) from a recipient at a later stage by use of hypnosis or administering sodium pentothal. Most of the binary data stream besides the tag(s) could simply have consisted of random data.

Purely as an example, if an alphanumeric tag was: HONEYBADGER1 it would appear when rendered in EBCDIC binary notation as:

11001000   11010110   11010101   11000101   11101000   11000010
11000001 11000100 11000111 11000101 11011001 11110001

And the same tag expressed in ASCII binary notation would be:

01001000   01011111   01001110   01000101   01011001   01000010
01000001 01000100 01000111 01000101 01010010 00110001

Every eight consecutive bits (as grouped above) make up a byte and this allows up to 256 alphanumeric characters and special characters to be encoded in each byte if using EBCDIC. With ASCII encoding only 7 of the 8 bits in each byte are used to specify the different characters and this system allows specification of 127 alphanumeric characters. So if, say, HONEYBADGER1 was written using ASCII encoding it would produce a similar but different string of binary ones and zeros. (Of course, any serious coder or person analyzing binary strings like this would have them printed in hexadecimal notation, rather than raw binary, making tag(s) easier to spot!)

An encoded label like this is rather like a barcode which can use as many as 43 ASCII characters to precisely identify a product. Today barcodes are instantly read using a barcode scanner in every grocery store in the land. Back in 1980 the same electronic technology was available but why would anyone want imprint an invisible marker on another human? If, say, US Special Forces men using a THW wanted to imprint some kind of identifying label on any person who touched the weapon, how could that have been achieved? And why might they need to do that?

When using non-lethal weapons to deal with terrorists—and maybe in particular hostage takers—it would certainly be useful to imprint some sort of secret marker on them if that were possible. If some miscreant was finally captured, say, years later and such a marker label could be physically scanned or extracted from his memory, it would show whether or not one had found the right person. When Osama bin Laden was located and shot dead by US Navy SEALs in Abbottabad in 2011 it was vital that he should be positively identified. In that case his DNA was taken for comparison with DNA previously supplied by some of his close relatives. So, from *Operation Neptune's Spear*, there was no doubt at all the USN SEALs had got their man.

The supposed string of binary digits was apparently somehow imprinted in Jim Penniston's memory when he touched the large triangle on the exterior of the UFO (which I suggest was the **A** for Apollo in NASA's Apollo Mission symbol). Whether he received the binary as a bright rapidly flashing beam of light or whether it was a digital electronic signal directly reaching his brain is not clear. He says he saw it in his mind's eye and he felt compelled to retain it in his head until he was able to write it down a few hours later. Most of us would probably think it impossible to remember a long string of binary digits and then to reproduce it accurately any time later. However some people who have what is called a photographic memory and have shown they are able to do just that with pages of text or numbers in every detail. It may be that this faculty is quite common but most people who have it are totally unaware of it. Photographic memory is most probably something that only operates in one's subconscious and, when such a "download" of information does occur, it may seem like the reception of a telepathic message.

If some US intelligence agency had developed such a psychotronic device by 1980 it would certainly have wanted to test it on some unsuspecting subject. The device was certainly separate from the DEW which apparently zapped Burroughs and one has to wonder whether such psychotronic weapons like this are in use today. Being able to imprint an identifying marker on enemy personnel or terrorists would certainly be a useful ability and this sort of test, if successful, would demonstrate the feasibility of transmitting such digital data or code to human targets.

It could well be the reason why Penniston was subjected to multiple interrogations using hypnosis and sodium pentothal "truth serum" soon after the RFI event. The NSA/DIA men who must have known what he had experienced in the forest were now trying to recover the binary code from him and so validate their new psychotronic weapon.

It certainly appears the interrogators failed to get what they were after. Penniston himself says that after feeling compelled to write down the long string of ones and zeros in his notebook he put it away in a drawer in his digs in Ipswich and forgot all about it. The pages of the notebook on which the binary code was written are dated 26th and 27th December 1980 and he insists the code was not added later. He is willing to have the notebook forensically tested to establish that—if that is scientifically possible. At the time, he apparently never saw there was any significance in the ones and zeros he had written down and he hadn't even heard of the term "binary code", or what it meant.

He says he never mentioned the binary ones and zeros he wrote in his notebook to the inquisitors, since it didn't seem relevant, and it may never have occurred to them they should search for a notebook.

Going back to what Kit Green stated about RF non-ionizing EM radiation being the cause of injury to John Burroughs, one has to ask whether Green knew, or suspected, that an American DEW was the source of it. If Green did suspect that, I believe he would not have revealed it. He would merely have said the damaging radiation was part of "the phenomenon" implying it was from an alien UFO rather than anything of US military origin.

Since the 1990s the USAF, and indeed the UK's MoD, have preferred using the more vague term UAP (Unidentified Aerial Phenomenon) rather than UFO. Expressed thus, their use of the term

"UAP" could therefore sometimes be used to cover up the source of any effects of secret US military testing. But whether or not a long string of binary ones and zeros was somehow downloaded into Jim Penniston's memory is something upon which we can only speculate.

Is it really possible to imprint strings of binary digit computer code in the mind of a human subject? It is certainly hard to believe but, even if that could be done, its interpretation as a New Age message from supposed extraterrestrial aliens appears nothing short of ridiculous!

PLASMA RELATED UAP AND PROJECT CONDIGN

Much is made in Nick Pope's 2014 book *Encounter in Rendlesham Forest* of the importance of *Project Condign* which was a highly classified intelligence study into the UFO mystery by the UK's MoD from 1993 to 1996. Nick Pope had worked in the Ministry of Defence in London for several years before he was appointed head of the MoD's UFO Project (or UFO desk) in 1991. The position was part of Sec(AS)—the Secretariat (Air Staff). Although the department acknowledged that at least 80% of UFO the reports it received could be explained as misidentification of ordinary objects such as aircraft lights, satellites, planets, stars, meteors, balloons and the like, it was conceded about 5% of reports still defied conventional explanation. Regularly the MoD would say these unknowns were of "no defence significance" and consequently the UFO Project was closed in 2009.

One of the most controversial aspects of the *Project Condign* study related to "plasma related fields" sometimes associated with UFOs, or UAP as MoD preferred. Whether any supposed UFO sometimes produced a plasma field or whether an observed UFO was in itself a plasma field was seldom clear. In any case, use of the term UAP avoided the, by then, loaded term "UFO" which certainly implied a physical craft which, in popular belief, might be thought to be of extraterrestrial origin.

Nick Pope was the chief proponent of commissioning a study of UAP for the MoD and the arbitrarily named *Project Condign* was highly classified as 'Secret UK Eyes A'. Study of UFO reports held by the MoD and UFO data it considered sensitive would be conducted by a nominated defense contractor with appropriate security clearances.

The resulting report found no evidence that UAP are hostile or under any type of control. It did not offer explanations for the phenomena nor did it deny or confirm that their origin might be extraterrestrial. It did however suggest that "plasma related fields" may have given rise to some of the reports of huge triangular craft in the sky. These are, of course, the UFT referred to earlier in this book which had been reported over parts of the US, over the UK, over Belgium and over several other countries. The issue of possible military application, or the possibility of weaponizing UAP then became a consideration.

At this point one might well ask what evidence did the writer(s) of the report have to maintain that plasma related fields *were* the UAP, or else produced the UAP that were under study? For years there had been occasional stories of witnesses being physiologically and/or psychologically affected by UFO encounters. Such stories included reported experiences of missing time and, indeed, some claims of alien contact and even alleged alien abduction.

However, there was no indication that Project Condign had accepted any of these anecdotal UFO stories as the reason for its suggestion that plasma had caused physiological and/or psychological effects in humans. Reading between the lines it's fairly plain that it was solely the Rendlesham Forest Incident itself which the report's compiler had in mind when describing "Non-ionizing EM effects on humans and "EM field from a plasma". The report declares that during the Rendlesham Forest event several observers *"were probably exposed to UAP radiation for longer than normal sighting periods. There may be other cases which remain unreported. It is clear the recipients of these effects are not aware that their behavior/ perception of what they are observing is being modified."*

I am certainly not suggesting that all UAP cases (a.k.a "UFO"s) are the results of secret military testing but it does look as if the writer(s) of the *Project Condign* report were proposing that UAP/ UFOs and their supposedly associated plasma fields were the explanation for what transpired in the Rendlesham Forest Incident. The truth of the matter that it must have been a secret US military test was either completely missed or else this explanation was deliberately fed to the MoD by US military authorities to cover up what they had done.

Proponents of UAPs being of extraterrestrial—or maybe even interdimensional—origin are suggesting that they are the product of some non-human intelligence rather than anything which is man-made. That is a thesis which is impossible to prove and it should be treated with the greatest caution. It is held to be the unstated position taken by some senior officials in the US military and also in the UK's MoD, but again the truth of that is impossible to determine. In any case, such belief as regards UAP is certainly no substitute for proof of a non-human origin.

The suggestion that UAP can result from "plasma related fields" is equally a rather dubious thesis. Yes, there are earthly manifestations of plasma as natural phenomena, such as ball lightning, but there is nothing to indicate such things are produced by an intelligence. Ball lightning remains an unexplained phenomenon and many scientists doubt its very existence. Wikipedia's current article on ball lightning describes an enigma similar to that of the UAP/UFO phenomenon but it doesn't provide any answers.

By invoking "plasma related fields" to describe some UAP, those who believe in an ET or interdimensional solution can also propose that many UAP observed are very likely holograms. Holograms are three-dimensional images formed by the interference of light beams from a laser or other coherent light source. They are artificially produced by humans and they are not known to occur in nature. Their connection with plasma is the fact that some very specific types of hologram result from focussing lasers to ionize multiple particles of air making an array of tiny points of light. Each of these points arranged in a 3-D pattern is actually a tiny bit of plasma.

Well, that may be, say those who believe UAP/UFOs are of alien origin. The aliens are most probably technologically advanced by hundreds, if not thousands, of years, they say, and so their holograms (which may spring from plasma related fields) are far far advanced compared with those produced by humans. With such advanced technology the aliens can obviously create Virtual Realities (VR) in front of us which are totally indistinguishable from our own physical reality and so appear as magic.

There is no answer to such ET speculation of course! Nevertheless, let's see these things for ourselves or at least find multiple reliable witnesses, not FMWC, whose experiences can be heard and tested.

RELEASE OF HOSTAGES & END OF PROJECT HONEY BADGER

If there were different kinds of US Trojan Horse Weapons being tested as part of *Project Honey Badger* by the 160th SOAR (Airborne)—or other US Special Forces—in December 1980, these never had to be used as part of a further US military attempt to free the hostages in Iran. Whether or not President-elect Ronald Reagan was briefed on the testing of the THWs being developed for Honey Badger in the final weeks of 1980 is a matter of speculation. Almost certainly he would have been briefed on the true story of the Rendlesham Forest Incident once he assumed the office of President in January 1981.

On January 20th 1981 at the moment President Reagan completed his 20-minute inaugural address after being sworn in at the Capitol in Washington DC, the 52 American hostages in Tehran were released to US personnel there. They were flown out of Iran, first to Algeria and then on to a US military base in Germany and to an Air Force hospital in Wiesbaden where they were received by ex-President Carter. Later they flew back to the US and ten days after their release were given a ticker tape parade through the Canyon of Heroes in New York City.

There are theories and conspiracy theories as to why Iran postponed release of the hostages until that moment in 1981 but some observers were of the opinion that people in Reagan's transition team were involved in secret negotiations with the Iranian government to secure the release. The Iranians would not allow any further dealing with the President Carter's administration and, it was rumored, they were finally persuaded to agree to the January 20 release in exchange for Reagan's promise to return hundreds of millions of dollars of Iranian assets that the Shah had retained control of outside Iran.

Like Project Honey Badger, details of what really transpired as regards the US agreement with Ayatollah Khomeini's revolutionary regime remain secret even to this day. And from a political angle very little has changed in Iran with the IRGC still the mainstay of the mullahs' repressive Shiite regime. The political slogan *"Death to America"* is still posted in public places and chanted by crowds of Iranian government supporters but at least the US administration, in April 2019, designated the IRGC as a Foreign Terrorist Organization.

In any case it's worth repeating here that development of special techniques as part of Project Honey Badger for use by US Special Forces during the second half of 1980 was officially acknowledged by USSOCOM several years later, as stated earlier:

> *Numerous special operations, applications, and techniques were developed which became part of the emerging US Special Operations Command repertoire.*

The testing of a THW equipped with a Directed-energy Weapon in Rendlesham Forest outside RAF Woodbridge was almost certainly intended to be a key element in Project Honey Badger. If the THW was going to be used immediately prior to a further attempt to rescue the hostages in early 1981, its purpose must have been to lure the leader of those guarding them out of the Teymur Bakhtiar mansion where the hostages were being held. Special Forces would then invade the mansion and rescue the hostages once the leader of the guards had been disabled or neutralized.

If that was the plan, it certainly explains why the CIA or NSA (or DIA) men who repeatedly and relentlessly interrogated Penniston and Burroughs soon after their ordeal in the forest were so anxious to know how effective their THW with its DEW had been. Not that they ever told the men what the weapon was, or what it was for. Until then this DEW was probably just an untested psychotronic weapon and for the next hostage rescue attempt to have any chance of success, the CIA had to know whether it would be effective. It is of course quite possible the CIA did not consider the exercise was a success and the Rendlesham THW would therefore not be used in any future action.

It was probably just as well that a dangerous second rescue operation was not attempted since the Iranian guards were expecting one and it might well have resulted in the killing of most of the hostages. After *Operation Eagle Claw*, the failed US rescue attempt on April 24 1980, the hostages were first moved to different prisons in Tehran in order to prevent further rescue attempts or escapes. However it was not until November 1980 that the Iranians moved all the hostages from various prisons and safe houses in the city to the Teymur Bakhtiar mansion on the outskirts of Tehran. Only then did feasible plans for a further rescue attempt by US Special Forces become possible.

"YOU CAN'T TELL THE PEOPLE"

An intriguing comment on the Rendlesham UFO affair came 16 years later from someone who was almost certainly briefed on what had really happened by senior UK government figures who must have demanded an explanation from US authorities. When news of the RFI came out in the press in 1983 the UK MoD had presumably been briefed about it but may not have been told the whole story. The comment "You can't tell the people" came from Baroness Thatcher—Margaret Thatcher, that is—who was Prime Minister of the UK from 1979 until 1990. During her premiership Thatcher was on extremely friendly terms with US President Ronald Reagan and it is very likely that he confided in her the true story of RFI.

Georgina Bruni produced a well researched a book on Rendlesham, *You Can't Tell the People*, which was published in 2000. She had attended a charity dinner in London three years earlier at which Lady Thatcher—who had just returned from Washington DC—was the guest of honor. Although Georgina Bruni didn't know Thatcher she went to speak to her at the end of dinner and she soon turned the conversation to UFOs and the Rendlesham affair. She asked whether Thatcher was aware that several US military officers had come forward in recent years with claims of an unusual nature, involving UFOs. She says she recounted her interviews with some military men and scientists who claimed they had worked with "alien technology".

That sounds to relate in particular to Col. Philip Corso whose claims in his 1997 book *The Day after Roswell* regarding alleged "alien technology" from the supposed Roswell UFO crash have largely been discounted. It seems very unlikely that Thatcher knew about Corso and his claims but she surely must have known something about what had happened in Rendlesham Forest in December 1980. Georgina Bruni says that she next asked Thatcher whether she would offer an opinion on UFOs and "alien technology" to which she received from her the reply that "You can't tell the people."

That of course became the title of Georgina Bruni's book which is all about RFI and the UFO saga there which began in December 1980. I'm sure that Margaret Thatcher did say that but I strongly suggest it was with regard to what had gone on at Rendlesham rather an answer to her question about UFOs and "alien technology". Bruni was a strong believer in an extraterrestrial UFO explanation and Margaret Thatcher's rejoinder was the ideal title for her forthcoming book in so far as it implied the ex-PM was well aware of the "alien presence".

Lady Thatcher continued "You must get your facts rights "and she then repeated "You must have the facts .... and you can't tell the people". No doubt she did say that—and it does sound like Margaret Thatcher's turn of phrase—but I hardly think she was talking about Philip Corso's cockamamie claims regarding "alien technology". She was surely inferring that what had happened in Rendlesham Forest just outside RAF Woodbridge in December 1980 was a secret US military test, almost certainly involving a secret weapon, and obviously something that could never be admitted or made public.

There are five specific reasons why certain agencies within the US government will never admit to what was implemented by them at that time and responsibility for the event now known as the RFI (Rendlesham Forest Incident). Before listing these reasons let me quote some of what Jim Penniston wrote in the 2014 book ERF saying that he and John Burroughs were still seeking answers to the question of why they were treated like enemy combatants by their fellow Americans during their RFI debriefing. What, he asks, could possibly justify this and what could possibly justify the psychological and, indeed, physical impact on their health? Penniston wrote:

> *I often think , so if my memory was messed with, then the obvious question is why? And if this was done by our own government, then there were things used for containment purposes, but then why would*

they allow the information about the binary codes and glyphs to exist in my notebook? I'm sure the information could have all been extracted under drugs or in the other parts of the interrogation, as described in my one and only hypnosis session. Then of course there is the recall of information going back to the landing and take off area at Rendlesham. Then other recall from one of the binary locations. All of this and then my question is, of course, why? The conduct of the agents while I was under the operational control of the USAF, makes me wonder why the Air Force would allow such things to be done to their personnel while on a base! Why would they, my fellow Americans, treat us as enemy combatants. Or even worse.

Jim Penniston goes on to say:

I do take medication that curtails the nightmares and flashbacks from that night. The diagnosis is PTSD, I suppose those things are just normal for me today. As I search for the answers and live with the debilitating effects from that night and follow up days, by my countrymen's interrogations, there are one or two questions that I want to ask and have answered. They are: "what was it that we saw and why have the United States and the United Kingdom done everything to cover it up as a UFO event?" and "What is it they are so afraid of the public finding out?"

He then adds that whatever it is they want to be kept so secret "obviously has nothing to do with UFOs [ETs or aliens] as is clearly shown in the declassified documents." Well, I believe that I can answer most of Jim Penniston's questions and most of the answers I suggest can be found in this book.

The main reasons why the US government, and their intelligence agencies which were involved, can never—even now—admit their role in the December 1980 Rendlesham Forest Incident are as follows:

(1) If the purpose and the technique embodied in this THW, which became known as the Rendlesham Forest UFO, had ever been made public it would never be of further operational use against an enemy.

(2) If the fact that a non-lethal psychotronic weapon such as a DEW had been tested on US servicemen without their prior knowledge and consent ever became known, it would be deemed unacceptable and something that was in breach of the US airmen's human rights.

(3) If this non-lethal weapon, which I call a THW, did include some kind of Directed-energy Weapon it would probably have contravened the CCWC (Convention on Certain Conventional Weapons) Treaty which was signed by the US and by many other countries earlier in 1980. (The treaty sought to prohibit or restrict use of certain conventional weapons which are considered excessively injurious or whose effects are indiscriminate. Although DEWs may not be listed specifically in the CCWC Treaty their use would certainly have violated the spirit of the treaty, if not the actual letter of it.)

(4) The clandestine carrying out of a test of a secret weapon on UK soil outside the perimeter of the RAF Woodbridge base (which was, together with RAF Bentwaters, under agreed US control) was almost certainly a violation of the NATO Status of Forces Agreement of 1951 (SOFA) and the Visiting Forces Act of 1952. Public disclosure of such a violation could have had serious political consequences for the NATO alliance and US/UK relations.

(5) The boilerplate Apollo CM (BP-1206) which appears to have been temporarily adapted as a THW during the Rendlesham incident is the property of the Smithsonian Institution and NASA. It was supposedly on loan to the USAF at the time. It seems unlikely that permission was given, or would have been given if requested, to adapt an Apollo CM for non-peaceful purposes outside the scope of NASA's mission.

If, indeed, it was Ronald Reagan who briefed Margaret Thatcher in 1981 on the test that US Special Forces had secretly carried out in Rendlesham Forest in December 1980, she would have certainly understood. She would have been completely aware of the desperate position that the US government had been faced with as a result of the Tehran Hostage Crisis and she would have been sympathetic. The Anglo-American Alliance, after all, was much more important than any other consideration. And, in not disclosing Reagan's confidence, her retort to anyone who asked her about the Rendlesham Forest UFO, would most probably have been "You can't tell the people".

Desperate circumstances sometimes require secret measures to deal with a desperate situation. Forty years on from 1980 the IRGC still exercises the same power in Iran and it has now been officially designated as a terrorist organization by the United States. It is small wonder that extraordinary measures were considered as going to be necessary in order to solve the hostage situation at that time.

CONCLUSIONS

The search for the truth about the Rendlesham Forest Incident (RFI) of December 26th 1980 and identification of that "UFO" which supposedly landed in the forest at the time has been a quest of mine that has lasted over twenty years. Most pieces of the extraordinary jigsaw have now been put into place and I'm confident that we know sufficient to offer informed speculation where additional explanation is required.

When awareness of the RFI entered the public domain in the 1980s it seemed to outsiders there were only two possibilities as to what this could have been. Either it really was some alien visitation by a UFO from outer space or else the US airmen, who had seen the craft and told of their experiences when they approached it in the forest, were delusionary or else they must all be fantasists and liars.

No third explanation seemed possible at the time.

Yet for years neither explanation rang true to me. During the 1980s and the 1990s around 50% of the US public said they believed in UFOs and their extraterrestrial origin and most thought aliens had visited or were still visiting this planet. Even today many think the US government knows a whole lot more about UFOs than has ever been admitted. And the 1980s were certainly an era when UFO belief was becoming something like a religion for many people with various stories of alien contact and even of human abduction by aliens.

If RFI had happened a hundred years earlier people might have suggested the visitors were ghosts or angels or even Santa Claus with his flying reindeer. The Hollywood movie *Close Encounters of the Third Kind* had gone on release in 1977 and the *Star Wars* movies made their debut in the same year. *ET* appeared in 1982. It led to huge public enthusiasm for the UFO subject.

Naturally many scientists and UFO skeptics were appalled by these new beliefs and sought more rational explanations for all this UFO nonsense as they saw it. But a general scientific disbelief in UFOs would persuade no one and the skeptics really had to disprove individual cases—especially high-profile ones like Rendlesham.

It was obviously easy enough to cast doubt on single witness cases of alleged UFO contact or alien contact since most such stories had no supporting evidence or independent corroboration. A few of the so-called contactees, like George Adamski and Billy Meier, had produced photographs of the supposed flying saucers flown by their alien friends but it didn't take much examination to show these were faked.

When it came to the RFI we were told that all the actual photos that had been taken were badly fogged and had been discarded as useless. Obviously that cast serious doubt on the whole affair but the UFO skeptics still had to explain the Halt Memo. That most did with quite a vengeance: the indentations in the ground made by the UFO they insisted were rabbit diggings and apparently raised radiation levels at the UFO landing site were really just normal background radiation.

Such explanations of the physical traces assumed the men involved were either of low intelligence or they were lying. Then came the other skeptical explanations, starting with the one that it could all be explained as misidentification of Orfordness lighthouse. The re-entry of Soviet satellite Cosmos 749, the truck load of burning fertilizer, the absurd story of British SAS men parachuting into the forest with black helium balloons to get their revenge on their USAF tormentors.

Again and again we were offered these skeptical explanations none of which was true. Something was wrong here as regards what the skeptics were maintaining and also wrong as regards unbelievable explanations the UFO was a visiting alien craft which had randomly come down in the forest close to an important American airbase.

What I did find very suspicious, and which many researchers ignored, was that the extraordinary events in Rendlesham Forest happened in the same week as the Cash-Landrum affair in Texas. Some people believed this was a completely separate UFO event or even, in view of the many US military helicopters that were accompanying it, that the UFO was an alien spacecraft being flown in some joint operation with the aliens. Not so, I thought! This and RFI were very likely part of the same military plan. But what was the purpose of the exercises?

Although my suspicions preceded the easy availability of Google and Wikipedia for research purposes it wasn't too difficult to read about the US government's desperate predicament that was being caused by the US hostage crisis in Tehran right through the year 1980. As we've seen, this led to a failed rescue attempt, *Operation Eagle Claw*, and the virtual unraveling of Jimmy Carter's presidency. Long after the return of the hostages when there was no further need for *Project Honey Badger*, it's existence was for the first time retrospectively acknowledged. Only then did it become apparent that there must have been an urgent need for a specialized THW to facilitate a new attempt to rescue the hostages. And when the timing of the RFI and that of the almost simultaneous Cash-Landrum Affair was checked, there could be no doubt that both of these events happened at just at the time when *Project Honey Badger* called for such weapons of deception. The timing in late 1980 was certainly no coincidence.

After Colonel Halt retired from the USAF in 1991 he made his first public appearance in a television documentary where he confirmed the authenticity of the RFI. In the following years he appeared in further TV documentaries and more recently released a book co-authored with John Hanson titled *The Halt Perspective*. Likewise Jim Penniston and John Burroughs both retired from active service during the 1990s (though Burroughs was for two years or so called back for active service following the 9/11 terrorist attacks) and subsequently both appeared in TV interviews and documentaries about the RFI. These two both made major contributions to Nick Pope's 2014 book *Encounter in Rendlesham Forest* in which they are jointly listed as co-authors.

The separate sketches of the UFO in the forest clearing made by each of these men and all the additional descriptions and information made it quite clear that the mysterious UFO in the forest had indeed been a real physical craft and that the facile suggestions made by the UFO skeptics were without merit. Yet the alternative to that which we had repeatedly been offered—that the UFO was of extraterrestrial origin—simply did not ring true either. The truth about RFI must, I thought, indicate there was a third possibility.

The breakthrough was, of course, the story of the sighting of that helicopter flying out from RAF Woodbridge late on the night of December 25th 1980 seen by the forester and his wife at Folly House. It was carrying a large conical object slung below it which could only have been a boilerplate Apollo CM. The helicopter could only have been one belonging to the 67th ARRS at RAF Woodbridge. The conical object must have been the "UFO" that was later approached by Jim Penniston and John Burroughs in the forest clearing. And BP-1206 with its secret history and its inadequately erased Apollo Mission symbol, is strikingly like what Penniston and Burroughs had sketched at the time, merely providing confirmation.

The cover-up operation as regards RFI was for years seen by the UFO community to be because senior USAF officers knew about the alien presence or possibly had some deal or alliance with the aliens. That would explain the secrecy and the evident need for containment. I say this is nonsense and the real need for the cover-up was because of the five considerations which I listed in the previous chapter.

The cover-up goes on to this day and there are still ex-USAF officers and others who will vehemently deny the facts that I've presented in this book. RFI was a top secret exercise that could never be admitted and the few who were in the know at the time—and also those who have been told privately years after the events—will likely disparage the story of the helicopter and its large conical load.

Having said that I'm quite sure there are USAF men and ex-USAF men alive today who are fully aware of the truth but will never admit to it. I'm told that ex-members of 67th ARRS and members of today's 67th Special Operations Squadron based at RAF Mildenhall, Suffolk, are most reluctant to discuss the matter. When questioned some have darkly suggested that if 67th ARRS pilots were involved in any way, these pilots may have been involved with drink or drugs at the time and that would account for any subsequent cover-up.

I don't think so! I do think the facts about RFI that are uncovered in this book are as close to the truth as we are likely to get. It is surely high time the truth was known and it's not for any political purpose that I think it should be brought out now. Nick Pope's book with John Burroughs and Jim Penniston, *Encounter in Rendlesham Forest*, (2014), was extensively based, we were told, on recently declassified government documents and is the only UFO book ever to have needed security clearance from both the British and American governments. I do not suggest that Nick is part of the Rendlesham cover-up that has gone on ever since 1980 but clearly there are UK and US government people, who were (or who are now) in the know as regards what really happened. They would still prefer you to believe it was an alien UFO rather than a secret US Special Forces exercise—which it was.

The alien UFO interpretation of RFI has now become part of British folklore. The Forestry Commission who manage Rendlesham Forest have set up a "UFO Trail" in the woods there and one can walk from the main car park by way of the disused East Gate to a small clearing where a supposed replica of the UFO (according to one of Penniston's sketches) can be found. The photo below shows me standing by it.

Above: The author standing beside the replica of the Rendlesham Forest UFO which can be found on the UFO trail.

It's not a very impressive or convincing ET UFO but, the Rendlesham Forest Incident has often been compared with America's Roswell Incident of 1947. That was when a supposed extraterrestrial UFO crashed near Roswell in New Mexico at the beginning of the flying saucer era. Over the years that story grew and grew with all sorts of alleged witnesses and UFO investigators jumping on board with claims of other alien spaceships crashing in the desert and the secret recovery of small alien bodies from the crash wreckage, both dead and alive. But, as was to be expected, the US military were said to have covered up or confiscated all the Roswell evidence and then falsely insisted the wreckage was merely an ordinary weather balloon.

From this great melange of claims, counterclaims and conspiracy theories the great Roswell UFO Myth was born. Today the town has a UFO Museum, has an annual UFO parade, and it attracts thousands of visitors every year, some of whom go on UFO tours of different alleged UFO crash sites.

Can the Rendlesham UFO myth ever compete with this? The replica UFO in the forest is a good start but I rather doubt a full scale UFO theme park is likely to follow. Interest in the UFO subject has waned and, sadly, no one has yet produced any concrete evidence at all of extraterrestrial visitation of this planet by aliens or, for that matter, any convincing remains of such aliens or their ET spacecraft.

I will say there is one fascinating parallel between the mysterious events at both Roswell and at Rendlesham. It is now almost certain that both were the result of top secret US military operations which were covered up with disinformation as soon as these events became publicly known. The crash debris found by Mac Brazel near Corona, NM, 75 miles NW of Roswell on July 7th, 1947 was that of a US Army Air Forces *Project Mogul* balloon-train previously launched from Alamogordo Army Airfield on June 4th. The Mogul program was a Top Secret attempt to detect sound waves from the first Soviet atomic tests. An array of microphones on the balloon-train 20 miles up in the stratosphere was used to listen for faint sounds from the Soviet nuclear test site 6650 miles away near Semipalatinsk in Central Asia. For those interested in what the 1947 Roswell Incident was really about, I would recommend reading Karl T. Pflock's book *Roswell: Inconvenient Facts and the will to Believe*, published in 2001.

Similarly, as outlined in this book, the Rendlesham Forest Incident was a Top Secret test by US Special Forces of a THW that was probably specifically devised for *Project Honey Badger*.

In attempting to unearth the whole truth about RFI I've been acutely aware that researchers must always evaluate accounts from the various witnesses that are inevitably going to be somewhat different. Whether or not the accounts of what they claim happened there in December 1980 are exactly as they now remember it, or whether we are being told embellished or, perhaps, incomplete versions of the truth is hard to assess. However, I should emphasize I do believe the three primary witnesses of the RFI are basically honest and sincere men who have recounted their experiences more or less as they remember them. However, they may well be holding back their suspicions of which particular agency was really responsible for their treatment and they are unwilling to point an accusing finger at some of their compatriots in the USAF or others in US intelligence.

I have little doubt that the story of RFI as presented in this book is a lot closer to the truth as regards what really happened there than any version of events previously published.

INDEX

P58